The Mathematical Laws of Natural Science

The Mathematical Laws of Natural Science
"unification is dead – long live unification!"

By
Keith Dixon-Roche

The Mathematical Laws of Natural Science

All concepts and formulas in this book not previously attributed to 'The Heroes' identified in Appendix A-4, are the sole property of Keith Dixon-Roche and protected by copyright.

Their use, publication, broadcasting, distribution, copying or any form of recording without Keith Dixon-Roche's written consent shall be a breach of international copyright law and subject to legal action.

Copyright © Keith Dixon-Roche 2017 to 2026

The Mathematical Laws of Natural Science

The Mathematical Laws of Natural Science
"unification is dead – long live unification!"

Published by CalQlata
info@CalQlata.com
First published February 2020
Final Publication June 2026

This book is sold subject to condition
that it shall not by way of trade or otherwise,
be lent, re-sold, hired out or otherwise circulated
without the publisher's prior consent and in such
circumstances it shall not be circulated in any form of
binding or cover other than that in which it is published
Copyright © CalQlata 2017 to 2026

The Mathematical Laws of Natural Science

The Mathematical Laws of Natural Science

Contents

1 INTRODUCTION 1
    1.1   ONE LAW FITS ALL: 2
    1.2   SCIENCE – A DEFINITION 3

2 ENERGY 5
    2.1   FUNDAMENTALS 8
        2.1.1   *Distance and Time* 9
        2.1.2   *Charge (electrical and magnetic)* 10
        2.1.3   *Temperature* 11
    2.2   FORCE 12
        2.2.1   *Potential* 13
        2.2.2   *Kinetic* 14
        2.2.3   *$E=mc^2$* 14
    2.3   ELECTRO-MAGNETIC ENERGY 15
        2.3.1   *Measurement* 17
        2.3.2   *Heat* 18
        2.3.3   *Light* 20
    2.4   STELLAR 22
    2.5   CALCULATIONS 23
        2.5.1   *Force* 24
        2.5.2   *Kinetic* 25
        2.5.3   *Potential* 26
        2.5.4   *Pressure* 27
        2.5.5   *Torque & Moments* 28
        2.5.6   *Power* 29
        2.5.7   *Field* 30
        2.5.8   *Electro-Magnetic (EME)* 31
        2.5.9   *Charge* 32
        2.5.10   *Microstates* 33
        2.5.11   *EME Deflection* 34
        2.5.12   *Temperature* 35
        2.5.13   *$E=mc^2$* 36

3 ELECTRICITY 37
    3.1   ELECTRICAL CHARGE 38
    3.2   DIRECT CURRENT (DC) 39
    3.3   ALTERNATING CURRENT (AC) 41
    3.4   CALCULATIONS 42
        3.4.1   *Resistivity* 43

The Mathematical Laws of Natural Science

4   MAGNETISM **45**

    4.1   MAGNETIC CHARGE *46*
    4.2   MASS *47*
        4.2.1   Constant @ Any Temperature *48*
    4.3   GRAVITY *49*
    4.4   CALCULATIONS *50*
        4.4.1   Bar Magnets *50*
        4.4.2   Solenoids and Coils *50*

5   ATOMIC PARTICLES **51**

    5.1   ELECTRON *52*
        5.1.1   Photon *53*
    5.2   PROTON *54*
        5.2.1   Hydrogen Atom *55*
        5.2.2   Hypothesis *55*
    5.3   NEUTRON *56*
        5.3.1   Neutron Creation *57*
        5.3.2   Neutron Energy *58*
        5.3.3   Neutron Energy Cell *59*
        5.3.4   Verification *60*
        5.3.5   Internal Stress *60*
        5.3.6   Hypothesis *60*

6   PROTON-ELECTRON PAIR **61**

    6.1   CALCULATIONS *64*

7   THE ATOM **67**

    7.1   ELECTRON SHELLS *69*
    7.2   NUCLEUS *71*
        7.2.1   Nucleic Structure *72*
        7.2.2   Nucleic Property Table *72*
    7.3   NEUTRONIC RATIO *76*
    7.4   ION *77*
    7.5   ISOTOPE *78*
    7.6   FUSION *79*
    7.7   FISSION *81*
    7.8   HALF-LIFE *83*
    7.9   CALCULATIONS *84*
        7.9.1   Atomic Density *84*
        7.9.2   Specific Heat Capacity *85*

The Mathematical Laws of Natural Science

8 THE STATES OF MATTER ... 87
- 8.1 LATTICE STRUCTURE ... 87
- 8.2 INTER-ATOMIC FORCES ... 88
- 8.3 VISCOUS MATTER ... 90
  - 8.3.1 Electron Clouding ... 90
  - 8.3.2 Surface Tension ... 91
  - 8.3.3 Elastic Moduli ... 91
  - 8.3.4 Viscosity ... 92
  - 8.3.5 Stress ... 92
  - 8.3.6 Electricity ... 93
  - 8.3.7 Magnetism ... 94
- 8.4 GASEOUS MATTER ... 95
  - 8.4.1 Partial Pressures ... 95
  - 8.4.2 PVRT ... 96
  - 8.4.3 Noble Gases ... 97
- 8.5 CALCULATIONS ... 98

9 ORBITS ... 101
- 9.1 TERMINOLOGY ... 102
- 9.2 ORBITAL LAWS ... 103
- 9.3 CENTRIFUGAL FORCE ... 104
  - 9.3.1 Circular Orbits ... 104
  - 9.3.2 Elliptical Orbits ... 105
- 9.4 LINEAR ORBITS ... 105
- 9.5 STATION KEEPING ... 106
- 9.6 ORBITAL PLANES ... 106
- 9.7 ORBITAL PRECESSION ... 107
- 9.8 CALCULATIONS ... 108
  - 9.8.1 Newton's Laws of Orbital Motion ... 108
  - 9.8.2 Centrifugal Force ... 110
  - 9.8.3 Orbits ... 112
  - 9.8.4 Linear Orbits ... 114
  - 9.8.5 Station-Keeping ... 115
  - 9.8.6 Orbital Precession ... 116
  - 9.8.7 Orbital Planes ... 117
  - 9.8.8 Planetary Mass ... 118

10 SPIN THEORY ... 119
- 10.1 CHICKEN & EGG ... 120
- 10.2 NO MOON ... 121
- 10.3 POLAR MOMENT OF INERTIA ... 122
- 10.4 INTERNAL HEAT ... 123

	10.5	MAGNETIC FIELD	124
	10.5.1	*Magnetic Reversal*	*125*
	10.6	GOODRICKE & ALGOL	125
	10.7	CALCULATIONS	126
	10.7.1	*Radial modifier (Δ) known*	*127*
	10.7.2	*Radial modifier (Δ) unknown*	*128*
	10.7.3	*Core Heat*	*129*
	10.7.4	*Magnetic Field*	*130*
11	**CORE PRESSURE**		**131**
	11.1	ACTIVE BODIES	131
	11.2	INACTIVE BODIES	132
	11.3	THE EARTH	133
	11.4	CALCULATIONS	136
	11.4.1	*Constant Density*	*136*
	11.4.2	*Variable Density*	*137*
	11.4.3	*The Structure of Celestial Bodies*	*138*
	11.4.4	*Celestial Fusion*	*138*
12	**THE UNIVERSE**		**141**
	12.1	THE AGE OF THE UNIVERSE	143
	12.2	THE BIG-BANG	145
	12.3	THE UNIVERSAL PROCESS	146
	12.4	CALCULATIONS	147
13	**CELESTIAL BODIES**		**149**
	13.1	THE ULTIMATE BODY	150
	13.2	THE GREAT ATTRACTOR	151
	13.3	GALACTIC FORCE-CENTRES	152
	13.4	STARS	153
	13.5	PLANETS	154
	13.5.1	*Inactive*	*154*
	13.5.2	*Active*	*154*
	13.5.3	*Gaseous*	*154*
	13.6	MOONS	155
	13.7	COMETS & METEORITES	155

14 THE MILKY WAY 157

- 14.1 HADES 158
- 14.2 OUR SUN 159
- 14.3 PLANETARY TEMPERATURE 161
- 14.4 MERCURY 162
- 14.5 VENUS 164
- 14.6 EARTH 166
- 14.7 MARS 168
- 14.8 ASTEROID BELT 171
- 14.9 JUPITER 172
- 14.10 SATURN 174
- 14.11 URANUS 178
- 14.12 NEPTUNE 180
- 14.13 PLUTO 182

15 THE PHYSICAL CONSTANTS 185

- 15.1 PRIMARY CONSTANTS 186
  - 15.1.1 *Principal Constants* 187
- 15.2 HEAT CAPACITIES 189
- 15.3 CHARGE CAPACITIES 190
- 15.4 UNITY 191
  - 15.4.1 *Principal Constants* 192
- 15.5 PHYSICAL CONSTANTS - EXPLAINED 194
  - 15.5.1 Σ 195
  - 15.5.2 ρ_u 196
  - 15.5.3 G 197
  - 15.5.4 φ 199
  - 15.5.5 $k, k', \mu_o, \varepsilon_o$ 200
  - 15.5.6 h, h' 202
  - 15.5.7 e, e' 206
  - 15.5.8 R_y, R_∞, a_o 207
  - 15.5.9 R_s 209
  - 15.5.10 R_n, t_n 210
  - 15.5.11 RAC, RAM 211
  - 15.5.12 N_A 212
  - 15.5.13 k_B, R_i 213
  - 15.5.14 $K \& h$ 214
  - 15.5.15 ξ_v, ξ_m 215
  - 15.5.16 B, RC 217
  - 15.5.17 X, X^R 218
  - 15.5.18 Y 220

The Mathematical Laws of Natural Science

A-1	GLOSSARY	222
A-2	REFERENCES	226
A-3	USEFUL FORMULAS	228
A-4	THE HEROES	229
A-5	NEWTON'S ORBITAL LAWS	230

    A-5.1 NICOLAUS COPERNICUS (1473 TO 1543) 230
    A-5.2 JOHANNES KEPLER (1571 TO 1630) 230
    A-5.3 GALILEI GALILEO (1564 TO 1642) 230
    A-5.4 ISAAC NEWTON (1642 TO 1727) 233
    A-5.5 PROOF (ELLIPTICAL ORBITS) 235
    A-5.6 EUCLIDEAN GEOMETRY (EQUAL AREAS) 236

        A-5.6.1 Proof (conservation of energy & equal time-swept area) 237
        A-5.6.2 Centripetal Force 241
        A-5.6.3 Distance Between a Satellite & its Force-Centre (R) 242
        A-5.6.4 The Inverse Square Law 243
        A-5.6.5 Orbital Period 244
        A-5.6.6 Constant of Proportionality 245
        A-5.6.7 Alternative Velocity Calculation 246
        A-5.6.8 Centrifugal force in an orbiting body 247
        A-5.6.9 Fundamental Laws of Orbital Motion 248

| A-6 | NEWTON ATOM VS PLANCK ATOM | 249 |
| A-7 | WHAT WENT WRONG? | 253 |

    A-7.1 THE ERROR 254
        A-7.1.1 Measured Vacuum 255
    A-7.2 THE PROBLEMS WITH RELATIVITY 256
        A-7.2.1 The Speed of Light 257
        A-7.2.2 Light Deflection 258
        A-7.2.3 Neutronic Radius (R_n) 259
        A-7.2.4 Station-Keeping 261
        A-7.2.5 $E=mc^2$ 263
        A-7.2.6 Hades 264
        A-7.2.7 Micro-Lensing (Event Horizon) 265
        A-7.2.8 PVRT 268
    A-7.3 QUANTUM THEORY 269

| A-8 | THE MOLECULE | 271 |

Preface

Michael Faraday once said; "*Nothing is too wonderful to be true, if it be consistent with the laws of nature*".

This scientific philosophy was abandoned at the beginning of the 20th century, since when innumerable fanciful and unprovable theories have been created to explain the laws of nature (astronomic and atomic) based upon a fundamental error; the photon.

The inability of scientists to explain the laws of nature based upon the photon has led to general acceptance of the phrase; "*the normal laws of physics do not apply*", may be used to justify any and all irreconcilable theories, which in turn, has led to the need for a unification theory to unite them. This is the reason why scientific advancement is still dependent upon experimentation, just as it was at the start of the industrial revolution.

The problem with this philosophy is that it facilitates the exploitation of ever failing projects based on flawed science. Because, if solutions are actually found for; pollution, waste, medicine, energy, transport, materials, chemistry, etc. this opportunity for never-ending profit would be lost.

As an engineer (not a scientist) I have been able to study the subject unconstrained by institutional dogma. Moreover, as an engineer, I understand the importance of accuracy and verification; as failure to do so can be catastrophic, which is not a constraint for scientists.

I have discovered that there is no need for a unification theory. A single unifying scientific theory that unites every branch of science in the universe; *energy,* was available to us as long ago as the 19th century.

The purpose of this book is to explain how the universe (and everything in it) actually works, in the hope that one day, somebody may have the will and the means to apply real science to the world's problems, for the benefit of all.

Newton and Faraday were not only visionary; they were also correct:

the laws of nature really are simple and consistent!

Keith Dixon-Roche

2020

The Mathematical Laws of Natural Science

The Mathematical Laws of Natural Science

1 Introduction

All of the Laws of Nature (without exception) must be:
    1) universal,
    2) necessary,
    3) invariable,
    4) simple,
    5) complementary,
    6) observable.
    7) provable.

The following are now indisputable facts of nature:

    1) everything in the universe is energy,
    2) mass is magnetic charge,
    3) gravity is magnetism,
    4) electricity and magnetism are essential converses,
    5) there are only 2 atomic particles in the universe; the electron and the proton,
    6) there are no such things as photons; light is electro-magnetic energy,
    7) there are no such things as; dark matter, black-holes, event horizons, sub-atomic particles, uncertainty, anti-matter, etc.

And finally:

Because all the laws of nature are now explained, scientific progress can and must be resumed for the benefit of all, not just the profiteering institutions.

Important Notes:

Mass is used here in this book only to avoid confusion for the reader. Whenever you see the term mass, you should read; non-polar magnetic charge.

Gravity is used here in this book only to avoid confusion for the reader. Whenever you see the term gravity, you should read; non-polar magnetic attraction.

The Mathematical Laws of Natural Science

1.1 One Law Fits All:

Numerous independent scientific fields and sub-fields have emerged over the last century, each of which generates its own nomenclature, formulas, units, laws, rules, etc. almost none of which interrelate, i.e.;

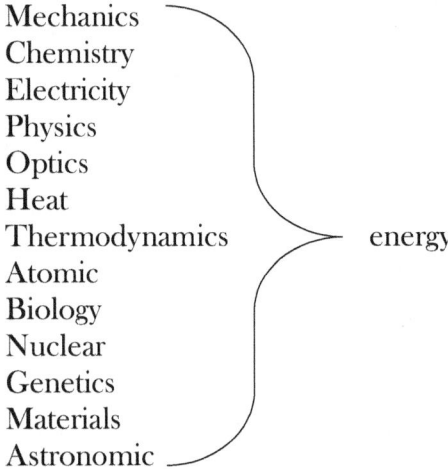

Mechanics
Chemistry
Electricity
Physics
Optics
Heat
Thermodynamics   energy
Atomic
Biology
Nuclear
Genetics
Materials
Astronomic

They are, however, simply different versions of the same entity; energy.

All of the above branches of science can be mathematically described using exactly the same formulas, exactly the same four constants, exactly the same two ratios and exactly the same units.

The above fields and sub-fields have evolved because the true nature of the atom and its particles, together with the true nature of all scientific constants, remain a mystery today.

But that should no longer be the case. All of physics can and should be united under one subject: The Mathematical Laws of Natural Science, which are all based upon energy.

Whilst it is understood that the above fields have their place in the world with regard to their use of these mathematical laws, the building blocks remain the same for all of them, for example:

Organic and inorganic chemistry apply to different forms of matter, but they both rely on the same proton-electron pairs, atoms and molecules, that react according to the same electrical and magnetic fields & charges.

In other words; whilst one may apply to the treatment of life and the other to the treatment of matter, the chemical reactions are identical.

1.2 Science – A Definition

Science is an all-inclusive mathematically verifiable description of all the natural features and events in our universe, from atoms to celestial bodies.

Conversely, a mathematical description that is only valid for a single universal natural feature or event, is not scientific.

That Isaac Newton understood this concept is self-evident from the title of his thesis; "Philosophiæ Naturalis Principia Mathematica" ("the mathematical laws of natural science"), which applies to all natural events and features from the atom to the 'Big-Bang'.

Note the term; "mathematica"; it means that all natural features and events must be explainable using [the same] mathematical laws - excluding statistics.

Irrespective of context or field, laws are fixed and all-inclusive; they are not open to interpretation or arbitration, nor are they selective.

Statistics relate to the consequences of laws, not the laws themselves.

Science is "knowledge", it is not "experimentation".

The correct scientific validation procedure is; first, develop the mathematical explanation, and only then, verify through observation.

Because this approach has been ignored (e.g. EHT, JET, Hadron collider, etc.) for more than a century, scientific progress is still based upon statistics, experimentation and guesswork, and is therefore no further advanced today than it was over a hundred years ago.

Any and all scientific theories that cannot be verified as such, must be rejected. These include; Relativity, Quantum Theory, uncertainty, dark matter, anti-matter, black-holes, sub-atomic particles, man-made global warming, CO_2 'the pollutant', event horizons, etc., none of which is representative of a single characteristic or event in nature. Moreover, any theory that relies for justification on the phrase *"the normal laws of physics do not apply"*, must also be rejected.

The Mathematical Laws of Natural Science

2 Energy

Everything in the universe is electrical and magnetic charges and the energy they generate.

The universe is an energy generator and storehouse. It generates energy via its celestial and atomic orbits which it stores in its neutrons.

Energy was not a concept known to Isaac Newton, so he used force to describe energy transfer; which is the manifestation of energy transferred between two or more bodies separated by physical distance.

Our universe comprises electrical and magnetic energy and nothing else; there is no such thing as mass or gravity.

Mass is a term of convenience for atomic particles - and collections thereof - that was derived for an unknown property, and gravity is the magnetic attraction between them.

Mass actually refers to packets of magnetic charge. Max Planck referred to these packets as Quanta, which is the term that will be used here to collectively describe the only two that are required to make the universe work; the proton and the electron.

Mass is therefore, the resistance to movement of Quanta that is under the influence of universal magnetism (gravity). Force is the effort required to overcome this resistance, energy is the amount of effort applied, and power is the rate at which energy is expended.

Apart from the non-polar magnetic charge present in all Quanta, electricity and magnetism are polar; negative or positive. Opposite poles attract and similar poles repel.

Because Quanta has opposite electrical polarity, when encountering other Quanta, their electro-magnetic energies will always be opposite or identical according to nature's requirements; i.e. polarity conflict is impossible.

The [orbital] kinetic and potential energy in all universal matter collectively equals the [neutron] energy released during a 'Big-Bang', and remains constant throughout the subsequent universal period.

All universal energy is generated by spin-friction within celestial satellites, stored in neutrons and released during subsequent 'Big-Bangs'.

The Mathematical Laws of Natural Science

All bright celestial bodies are satellites with sufficient sub-satellite mass that generate fission in their core atoms through planetary spin. Satellites with no force-centre or sub-satellites generate negligible EME. EME generated and emitted by proton-electron pairs that is not trapped, will be lost from the neutron energy released by the subsequent 'Big-Bang'.

The entire universe comprises a fixed, unchanging quantity of energy (7.4E+60 J), it always has done and always will. It was originally released by 3% of the neutrons in the ultimate-body, released during the last 'Big-Bang'; it remains unchanged today.
[first law of thermodynamics]

Environmental EME is trapped by orbiting electrons, which they convert into kinetic energy, and instantaneously - together with their proton partners - generate EME of their own, that they radiate back into the environment. This is the mechanism by which energy is transferred between atoms. Generated EME, rises and falls with the energy in the environmental EME.
[second law of thermodynamics]

The natural (minimum entropy) state of the universe is the collection of all Quanta into a single entity through magnetism; the ultimate-body.
[third law of thermodynamics].

The fundamentals of all forms of energy are; electrical charge (e), magnetic charge (m), time (t) and their relative distance (d), that together constitute the only four properties required to describe every form of energy in every branch of science; $kg.m^2/s^2$.

Note: electrical kinetic and potential energies should be calculated thus; $KE = ½.e.v^2$ $\{C.m^2/s^2\}$ & $PE = e.a.d$ $\{C.m^2/s^2\}$. But until electrical and magnetic charges are adopted correctly, we will continue to convert electrical charge to magnetic charge in order to define force and energy.

The Mathematical Laws of Natural Science

We perceive these energies like this:

> Mass is the resistance we feel in trying to move or deviate viscous matter.
>
> Gravity is the magnetic force we feel when next to a massive body.
>
> Light is a range of electro-magnetic energy that we can see with our eyes.
>
> Heat is the electro-magnetic energy we feel through our atoms.
>
> Viscous matter is that which we can handle.
>
> Gaseous matter is that which we breathe and keeps us warm.

Low-temperature scenario: *when you see an object, such as a cup, you are seeing all the adjacent atoms in that cup held together with magnetic field energy. In this form, the atoms are sufficiently close together to prevent the atoms in, say, your hand, from passing between the atoms in the cup, allowing you to touch but not penetrate the cup.*

The weight you feel when you lift the cup, is created by the magnetic energy between the Quanta in the cup and those in the earth.

High-temperature scenario: *If sufficient electro-magnetic energy (heat) is trapped by the electrons in the cup and your hand, the electrical charge energy in all the protons will exceed the magnetic field energy forcing the atoms in the cup and your hand to repel each other and intermingle, in a form that we understand as gas.*

Energy cannot be lost or gained, and it can only be transferred by electro-magnetic radiation.

2.1 Fundamentals

As defined in chapter 13.1, there are only seven fundamental constituents that may be used to mathematically define all forms of energy (chapter 1.1):

distance (d), time (t), electrical charge (e), magnetic charge (m)

The neutronic values of which are:

distance (neutronic radius); R_n = 2.81793795383896E-15 m

time (neutronic period); t_n = 5.90596121302193E-23 s

electrical charge (electron); e = 1.60217648753E-19 C

magnetic charge (electron); m_e = 9.1093897E-31 kg

The fundamental density of matter is that of an atomic particle:

ρ_u = 7.12660796350450E+16 kg/m³

number of electrons in ρ_u; ϵ_e = 7.82336489952175E+46

number of protons in ρ_u; ϵ_p = 4.26074122343073E+43

The universal constant Σ is the product of their reciprocals (chapter 15.5.1):

$\Sigma = 1 / \epsilon_e . \epsilon_p$ = 3.0E-91 (exact)

The fundamental ratios are as follows:

static; $\xi_m = m_p/m_e$ = 1836.15115053207

dynamic; $\xi_v = c/v_o$ = 1722.0458764934

The Mathematical Laws of Natural Science

2.1.1 Distance and Time

Distance and time are two of the only four units of measurement required to describe every property and characteristic in all branches of science.

Distance is the length of path between two points; either straight or circuitous, and time is the passage of time between events. They are universal constants; they do not deform around massive celestial bodies.

Distance and time define velocity ($v = d/t$) and acceleration ($a = d/t^2$).
The metric and Imperial unit for the measurement of time is the second.
The metric unit for the measurement of distance is the metre, and the Imperial equivalent is the foot.

Velocity is used to define kinetic energy.
Acceleration is used to define potential energy.

2.1.2 Charge (electrical and magnetic)

Electrical and magnetic charge are two of the only four units of measurement required to describe every property and characteristic in all branches of science.

Electrical and magnetic charges are the properties of the only two atomic particles that exist in the universe, the electron and the proton, working as an orbital pair. Rotate a negative electro-magnetic charge about a positive electro-magnetic charge and you will generate, and radiate, electro-magnetic energy (EME) of the same magnitude as the electron's kinetic energy.

The relative magnitude of electrical and magnetic charge is the relative charge capacity (RC); Coulombs per kilogram.

Electrical and magnetic charges radiate constant force fields throughout the universe. Whilst these fields are constant, irrespective of distance, they are distributed over the spherical area at that distance; $A = 4\pi.d^2$

As a proton-electron pair, the orbiting electron partner collects EME from its environment, which it converts to kinetic energy; altering its orbital velocity, and transfers to its proton partner.

The Mathematical Laws of Natural Science

2.1.3 Temperature

Temperature is a measurement of convenience for the kinetic energy of an electron in a proton-electron pair.

The measured temperature of an atom is that of the proton-electron pairs whose electrons are orbiting in shell-1.

The temperature of outer-space is that generated and radiated by all celestial bodies.

There is no such thing as 'zero' temperature.

Temperature is a term of convenience we use to measure the energy of the EME emitted by a proton-electron pair. It was configured by the following scientists in the 19th century:

Kelvin: $\underline{T}_K = PE / k_B \cdot Y$ {K}

Celsius: $\underline{T}_C = \underline{T}_K - 273.15$ {°C}

Rankine: $\underline{T}_R = 9/5 \times \underline{T}_K$ {R}

Fahrenheit: $\underline{T}_F = \underline{T}_R - 459.67$ {°F}

Because temperature is calculated thus:

$R = X^R / \underline{T}$

@ $\underline{T} = 0K$ the electron's orbital radius will be infinite (R=∞)

An orbiting electron will have left its proton partner long before this occurs. It can therefore be safely stated that

there is no such thing as zero temperature.

2.2 Force

A mass under constant velocity possesses kinetic energy, but induces no force.

A force is the influence of a mass due to acceleration.

Potential energy is the distance over which this influence is applied or occurs.

There is only one force in the universe; potential. However, this may appear a little confusing, because '*force equals mass times acceleration*' ($F = m.a$), there appears to be relative movement between the bodies, as implied by the acceleration.

But this acceleration could be static; due to magnetic charge;

$$g = G.m/d^2 \quad \{m/s^2\}$$

or dynamic, due to a change in velocity;

$$a = d/t^2 \quad \{m/s^2\}$$

There is no force due to kinetic energy, until the associated velocity changes, which requires an induced acceleration, when the force becomes potential. This is why the term 'KERS' used by vehicle manufactures is actually incorrect:

KERS = kinetic energy recovery system

But the energy recovered does not come from the vehicle's kinetic energy, recovery only occurs when the brakes are applied; deceleration. I.e. the system is recovering potential energy, not kinetic energy. The correct term should be:

PERS = potential energy recovery system

The energy is lost when the vehicle accelerates, and is recovered under deceleration.

2.2.1 Potential

Potential refers to the straight-line (linear) influence of a force on a body that may be induced either by a charge (electrical or magnetic), field (electrical or magnetic), or directly (physically) such as in a torque or moment.

Potential energy is the distance over which a force is applied:

$$PE = F.d \quad \{J\}$$

Potential energy is the negative (or positive), attraction (or repulsion) between all of the Quanta in the universe, irrespective of separation distance (d).

What we currently refer to as gravity is the magnetic potential energy between all universal Quanta due to their non-polar magnetic charge. Isaac Newton gave us the formula for this gravitational force as follows:

$$F = G.m_1.m_2/d^2 \quad \{N\}$$

But this formula implies that the force diminishes with the square of the distance between m_1 & m_2, which cannot be true, as it contravenes a fundamental law of nature; *the conservation of energy.*

Whilst the potential energies radiated by both magnetic and electrical charges retain their magnitude irrespective of distance, they are distributed over the spherical area at that distance. This is why such a potential force between any two bodies (gravity) *appears* to diminish with the square of the distance between them.

His formula should therefore look like this:

$$F = G.m_1.m_2 / 4\pi d^2 \quad \{N\}$$

And his constant should look like this:

$$G = 4\pi.a.c^2/m_u = 8.38628344228055\text{E-}10 \text{ m}^3 / \text{kg.s}^2$$

The Mathematical Laws of Natural Science

2.2.2   Kinetic

A free-moving body (in a vacuum) exerts no force on anything, until it comes under the influence of matter (viscous or gaseous), due either to impact or field (electrical or magnetic), whereupon this force will become potential.

Kinetic [dynamic] energy is the intrinsic energy in a linear (or curvilinear) moving body, Its mathematical relationship is:

$$KE = \tfrac{1}{2}.m.v^2 \qquad \{J\}$$

where: 'm' is the mass of the body and 'v' is its velocity

Kinetic energy, which exists in all moving particles, is always positive and induced via electrical, magnetic, electro-magnetic or impact (potential energy).

Kinetic energy in a satellite following a circular orbit (such as in a proton-electron pair) is not induced into the satellite by its force-centre such as in elliptical orbits; it must be provided by the satellite itself.

2.2.3 $E=mc^2$

Henry Poincaré's relationship between potential and kinetic energies is explained in chapter 2.5.13, but it essentially applies to circular orbits (e.g. the proton-electron pair) and works like this:

The potential energy between a satellite and its force-centre in circular orbits is always exactly twice the satellite's kinetic energy;

$$PE = 2.KE \qquad \{J\}$$

$$PE = 2 . \tfrac{1}{2}.m.v^2 = m.v^2 \qquad \{J\}$$

At the '*speed of light*' (c), the formula becomes:

$$PE = m.c^2 \qquad \{J\}$$

2.3 Electro-Magnetic Energy

EME possesses no mass; it travels at 'light-speed' irrespective of medium.

EME is only radiated by proton-electron pairs.

EME is the electro-magnetic field energy generated and radiated by rotating an electro-magnetic charge about another. Because it is generated by a proton-electron pair, its shape will be a helix, varying between plus and minus electrical energy and minus and plus magnetic energy.

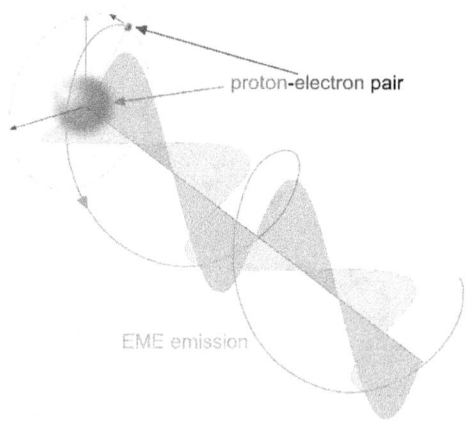

Whilst its wavelength, amplitude and frequency vary with temperature, it always travels at exactly the same velocity, which we refer to as light-speed (c), but is of course, the same speed for all EME.

EME is emitted in a direction normal to the plane of a proton-electron pair according to the 'right-hand-rule', and will travel in a straight line until deflected by reflection, diffraction or magnetic charge.

EME wavelength (λ), amplitude (A) and frequency (f) are responsible for the electro-magnetic spectrum (light, radio, X, γ, etc.), which is defined by an orbiting electron's kinetic energy. Its intensity (e.g. brightness) is defined by the concentration of emanations. Calcium, for example, is an atom with twenty proton-electron pairs and will therefore emit EME in twenty directions. Millions of such atoms will emit EME in millions of directions.

The EME generated by a proton-electron pair rises and falls with the kinetic energy (velocity) in the orbiting electron, which rises and falls with the energy in the radiation feeding it. In this way, [heat] energy in matter naturally stabilises with the energy in its environment.

If a proton-electron pair falls below temperature $\underline{T}_x$, the proton will no longer be able to hold onto its electron partner, which will fly off on its own at 17162.242521927m/s. This pair will no longer radiate EME; they will exist as lone particles. However, the EME in outer space is at ≈2.7255 K, so, $\underline{T}_\chi$ is not possible in our universe, even in outer space.

On reaching temperature $\underline{T}_n$, the orbiting electron's velocity will reach light-speed (c). Once this is achieved, the electron will combine with its proton partner to create a neutron. This pair will no longer radiate EME, but it will store the energy it was generating at the moment of unity.

The kinetic energy of an electron in free-flight will be that which it had when it was released from a proton partner; it will not alter with EME. Electrons in free-flight cannot collect EME because they cannot pass it on.

The Mathematical Laws of Natural Science

2.3.1 Measurement

The two definitions for EME energy are Boltzmann and SHC, but they appear to generate different temperatures, for example, @ 300K;

Boltzmann: $E_B = k_B.Y.\underline{T} = 3.94042969432577E\text{-}20$ J

SHC: $E_s = SHC.Y.m_m.\underline{T} = 1.97021484716289E\text{-}20$ J

$E_B = 2.E_s$!

However, EME has two measurable energies; *magnitude* and *range*.

Magnitude is a measure of the *kinetic* energy (KE) in the orbiting electron generating the EME, and is used in the calculation of specific heat capacity; SHC {J / K.kg}, where 'J' refers to EME *magnitude*.

Range is a measure of the *potential* energy (PE) in proton-electron pair that defines electro-magnetic energy (EME) using Boltzmann's constant; k_B {J/K}, where 'J' refers to EME *range*:

$|KE^-| + |KE^+| = 2.KE = PE$

which is equal to the *potential* energy in the proton-electron pair generating the EME.

When quoting EME energy generated, it is usual to specify its *magnitude*, rather than its *range*, but they both equally apply.

I.e., both calculations are correct because they give the same temperature:

$\underline{T} = KE / SHC.Y.m_n = PE / Y.k_B = 300$ K

where:
KE = the orbiting electron's kinetic energy
SHC = specific heat capacity of a proton-electron pair
Y = heat transfer constant
m_n = the mass of the pair ($m_e + m_p$)
PE = the potential energy in the pair (PE = 2.KE)
k_B = Boltzmann's constant

2.3.2 Heat

Heat is the EME generated by the kinetic energy in an atom's electrons. The heat we *feel* is from the senses we have developed to tell us when this kinetic energy is too high or too low. Electron kinetic energy is generated by the EME it absorbs from its surroundings.

It is important to remember that all the EME generated in the universe is just that; EME. It possesses; no light or heat – nothing, apart from energy, and the entire spectral band is simply a single range of EME from 3.17665E+09Hz to 1.6932E+22Hz. We [humans] have split the spectrum into special bands; "γ, X, ultra-violet, light, infra-red, micro, radio" for our own convenience, these bands will mean nothing to any other form of life.

If you or I, devoid of electrons - impossible I know, but bear with me - were to sit in the space between the sun and the earth, we would not be able to detect the sun's radiated EME. It would be invisible in every sense to the fictitious you (or me). EME is useless to all forms of life unless it can be detected.

Whilst EME doesn't deteriorate with distance travelled, we don't feel the sun's surface temperature (5788K) here on Earth because the energy *density* (Joules per square metre) radiated at the sun's surface is distributed over a spherical surface area between 45,000 and 48,000 times greater (dependent on the time of year). Therefore, the EME *density* we receive will be correspondingly less.

Life here on Earth, has evolved to detect and use this energy through our complex molecules. The trouble is, such molecules have energy tolerance levels, outside which they would no longer function; i.e. their state-of-matter, strength or condition (gas-viscous) could change, or inter-atomic bonding could fail.

If a block of viscous iron, the function of which is to be solid, received sufficient EME to increase its proton electrical charge energy above that of its atomic magnetic field energy, it would become a gas. And it would cease to be *a block of iron*; i.e. no longer functional.

You can't damage a block of iron, so it doesn't need senses. It doesn't matter how many times you change it from gas to viscous and back again it always returns to iron. The higher its temperature the stronger its atoms remain, until the innermost electrons achieve the *speed of light*, when it will become a different element (Z-1 or Z-2).

We (humans) have five senses – if you exclude time – smell, touch, taste, sight and hearing; each of which were developed for use and protection.

There are tolerance levels regarding acceptable amounts of EME any living organism can receive and remain functional. Therefore, all living organisms have developed senses, that can be used to ensure that these tolerance levels are maintained.

We can now conclude that **all heat is radiated**, even within viscous matter. Conduction is simply a term used to describe heat transfer within viscous matter.

Moreover, **convection** is not the transfer of heat upwards; heat is radiated equally in all directions. Convection is simply the highest temperature atoms - i.e. those with the greatest repulsive e' - are simply pushing away neighbouring atoms. This will inevitably cause them to move upwards; away from a planet's magnetic attraction (gravity).

All the EME in our environment is shared between all of our electrons. The greater the EME *density*, the greater the *heat* we feel. Irrespective of the *temperature* of the atoms that generated the EME, if the energy *density* can be shared throughout all the electrons in our body without exceeding its tolerance levels, we will remain functional. For example:

200,000 calcium atoms at 5788K will radiate 4.453E-13 J of EME.

A body of 10 iron atoms with a surface area 100th that of the calcium atoms, 100 metres away would absorb a total of 3.546E-20 J of heat energy, resulting in a temperature rise of 9K in the iron.

In other words; it is not the *temperature* of the atoms emitting the EME that defines our body-temperature; it is the quantity of EME absorbed by our body's electrons.

The Mathematical Laws of Natural Science

2.3.3 Light

We humans have designated light as EME with a wavelength ranging between 4E-07m (blue) and 8E-07m (red). We (humans) only call it light because we can *see* it with an unaided eye.

Ultra-violet and infra-red are also EME ranges, but we don't call them light because we can't *see* such wavelengths without artificial aids.

Question: *"If colour is defined by EME, and the surface of the sun is at a temperature of 5788K and looks yellow; how can my towel (at 300K) also look yellow?"*

Colour is a range of EME wavelengths that we cannot detect until it is absorbed by the electrons in our optical receptors (eyes).

Light is the intensity of visible EME wavelengths, or put simply; the number of EME rays per square metre.

We (humans) have developed eyes to detect a particular bandwidth of EME that best suits our purpose. Other lifeforms have developed receptors that best suit their own environment that may be outside the aforementioned *optical* range. Irrespective of a lifeform's preferred optical bandwidth, the purpose of sight is the same; to see what's in its environment and how best to exploit it.

Every proton-electron pair in the universe, *including that block of wood in the garden*, emits EME at a wavelength commensurate with the kinetic energy in its electrons. But you cannot see the EME radiated by that block of wood here on Earth, because it is radiated in the infra-red range. This is the reason infra-red cameras reveal objects in the dark here on Earth. They are actually capturing the electro-magnetic energy given off by the objects themselves, not the electro-magnetic energy radiated by the sun.

Unlike in a prism, the diffraction of light through natural matter is not organised. The image of that block of wood, is the sun's EME reflected and/or refracted by or through its constituent atoms and molecules.

Just as with heat, if the EME received by your eyes is not too intense, i.e. its density remains within your eye's tolerance levels, the sun's rays you see will do you no harm.

The Mathematical Laws of Natural Science

Whilst our sun's atmosphere comprises proton-electron pairs as hydrogen and helium heated to 5788K by fissionable energy, its internals comprise all of nature's elements, the proton-electron pairs of which give off EME according to their atomic shell numbers. A bright star therefore emits a preponderance of electro-magnetic radiation at the atomic temperatures from shell-1 (highest) to shell-N (lowest). But its internal matter is radiating EME from all of its electron shells @ temperatures from $\underline{T}_n$ to 125K, which is why distant stars look white; all the colours of the electro-magnetic spectrum are travelling towards us along the same path.

So, when you see a *yellow* towel here on earth, your eyes are detecting the EME radiated by the sun's outer atmosphere at a wavelength of ≈6.3E-07m diffracted by the molecules in your towel, but at an intensity that will not harm your eyes.

The Mathematical Laws of Natural Science

2.4 Stellar

Bright stars generate and store almost (>99.9%) all universal energy.

All bright stars (and planets) are simply gas satellites that have collected sufficient sub-satellite mass to generate fissionable energy, through spin-friction, in their core atoms, creating neutrons and radiating EME into universal space.

Because lone protons (H^+) have no way of absorbing or emitting EME (heat and light), the surface atmosphere of our sun must be proton-electron pairs (H), the surface temperature of which is said to be about 5788K.

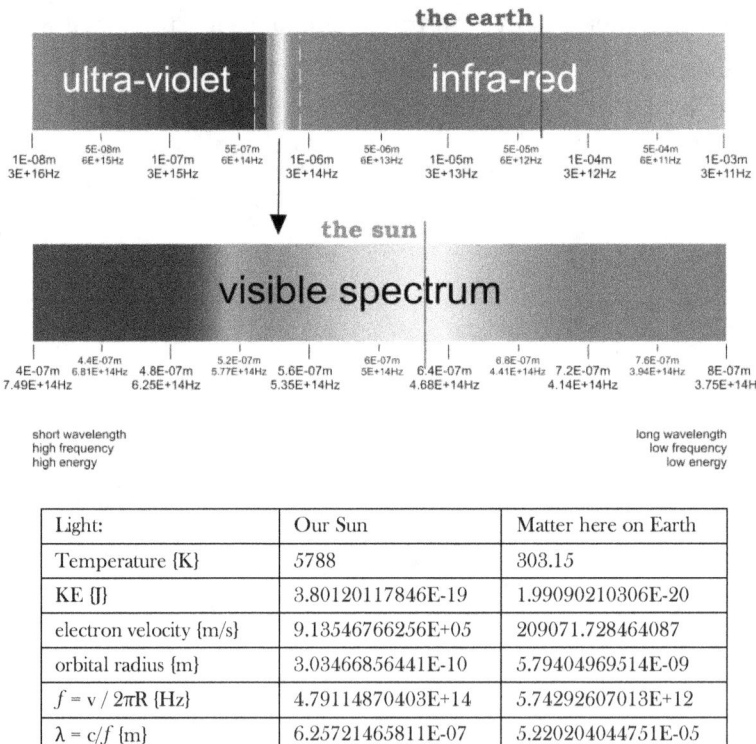

Light:	Our Sun	Matter here on Earth
Temperature {K}	5788	303.15
KE {J}	3.80120117846E-19	1.99090210306E-20
electron velocity {m/s}	9.13546766256E+05	209071.728464087
orbital radius {m}	3.03466856441E-10	5.79404969514E-09
$f = v / 2\pi R$ {Hz}	4.79114870403E+14	5.74292607013E+12
$\lambda = c/f$ {m}	6.25721465811E-07	5.220204044751E-05

Yet, our sun's colour and the infra-red light radiated by matter here on Earth can be predicted using Newton's atomic model ...

... demonstrating that the hydrogen at the surface of the sun is predominantly proton-electron pairs, which in turn means that these hydrogen atoms (proton-electron pairs) *must be a by-product of fission.*

2.5 Calculations

This chapter provides the formulas you may need to perform calculations for all forms of energy.

2.5.1 Force

Force is the effort required to induce acceleration in quanta, either by changing their velocity or direction:

$$F = m.a \qquad \{kg.m/s^2\}$$

Gilbert generated the original (and fundamental) formula for calculating the force between two bodies of quanta:

$$F = K.v_1.v_2/d^2 \qquad \{kg.m/s^2\{$$

'K' is a constant, 'v_1' & 'v_2' are variables and 'd' is the distance between them

Newton used this method to calculate magnetic force (gravity):

$$F_m = G.m_1.m_2/d^2 \qquad \{kg.m/s^2\}$$

because magnetism is accrued, m_1 & m_2 can be any value

& Coulomb used this method to calculate electrical force:

$$F_e = k.e_1.e_2/d^2 \qquad \{kg.m/s^2\}$$

because electricity is shared, 'e' must be the least value of the two charges; $F_e = k.(e/d)^2$

The ratio between the two forces (magnetic and electric) is the coupling ratio:

$$\varphi = F_m/F_e = 4.40742111792334\text{E-}40$$

Joseph Henry used a different method to calculate magnetic force:

$$F = \mu . (2\pi.I)^2 \qquad \{kg.m/s^2\}$$

$\mu = m.R/e^2 \qquad \{kg.m/C^2\}$
I = electrical current (intrinsic or induced)
m = mass (magnetic charge)
R is the distance between quanta
E is the electrical charge

When applied to the proton-electron pair in the neutronic Condition ...

$$E_A = \mu_o.(e/t_n)^2 . (2\pi)^2 . R_n^3 = 6.50120024835403\text{E-}43 \quad \{kg.m^4/s^2 = J.m^2\}$$
$$E_A = m_e.R_n/e^2 . e^2/t_n^2 . (2\pi)^2 . R_n^3 \qquad \{kg.m^4/s^2 = J.m^2\}$$
$$E_A = m_e.R_n^4 . (2\pi/t_n)^2 \qquad \{kg.m^4/s^2 = J.m^2\}$$
$$E_A = m_e.R_n^2 . c^2 \qquad \{kg.m^4/s^2 = J.m^2\}$$
$$E_A = h_e^2.m_e = 6.50120024835404\text{E-}43 \qquad \{kg.m^4/s^2 = J.m^2\}$$

... i.e. Joseph Henry's force is related to Isaac Newton's constant of motion.

Moreover, when we divide this value by volume (R_n^3), we get neutronic force;

$$F_n = h_e^2.m_e/R_n^3 = 29.0535538991261 \text{ kg.m/s}^2$$

We can use these force formulas to calculate the forces exerted by magnets.

The Mathematical Laws of Natural Science

2.5.2 Kinetic

The kinetic energy of a satellite orbiting in a circular path is exactly half the satellite's potential energy: $KE = \frac{1}{2} \cdot PE$

Kinetic energy can also be applied to angular movement (spin).

Spin energy in the proton:
$$SE_p = KE_e + PE = \tfrac{1}{2} \cdot m_e \cdot v^2 + (-)m_e \cdot v^2 \quad \{J\}$$
$$SE_p = -\tfrac{1}{2} \cdot m_e \cdot v^2 \quad \{J\}$$

Kinetic energy in the electron:
$$KE_e = \tfrac{1}{2} \cdot m_e \cdot v^2 \quad \{J\}$$

The total kinetic energy in a proton electron pair is therefore:
$$E = SE_p + KE_e \quad \{J\}$$
$$E = 0$$

[first law of thermodynamics]

2.5.3 Potential

Its linear mathematical relationship is: $PE = m.a.d$ {J}
in which 'a' may be positive or negative.

In a balanced system (e.g. orbits), both PE and CE must be equal. And in circular orbits, such as atomic, potential energy between Quanta is always twice the kinetic energy in the orbiting satellite, so their mathematical relationship is:

$$PE = 2.KE = 2 \cdot \tfrac{1}{2}.m.v^2 = m.v^2 \qquad \{J\}$$

2.5.4 Pressure

Pressure, in both states of matter, acts in three dimensions; it is actually force-density, as opposed to mass-density. But we almost always need to consider pressure in one direction, making it force per unit area, instead of force per unit volume.

If we multiply force per unit volume by a unit length, we get:

$$p \; \{N/m^3\} \cdot 1 \; \{m\} = p \qquad \{N/m^2\}$$

it is now in a form we can use.

The pressure in matter, irrespective of its state (gaseous and viscous), may be calculated either by the universally known and accepted PVRT formula ...

$$p = R_i/(RAM.N) \cdot \underline{T} \cdot \rho = R_a/N \cdot \underline{T} \cdot \rho \quad \{N/m^2\}$$

<small>where 'N' is the number of atoms in the molecule (1 or 2).</small>

... or like this:

$$p = -PE_1/Y \cdot \rho/m_M = k_B \cdot \underline{T} \cdot \rho/m_M \qquad \{N/m^2\}$$

<small>where 'ρ' is the density of the matter 'm_M' is the molecular mass.</small>

all of which give exactly the same results.

2.5.5 Torque & Moments

Torque and moment are both simply mechanical forms of energy:

$T = M = F.d$ {N.m}

where 'd' is the distance at which the force is applied.

It is a means whereby a fixed input force can be made to alter an output force, simply by changing the relative distances at which the forces are applied. They are both effectively levers.

The Mathematical Laws of Natural Science

2.5.6 Power

Magnetic power is the rate of magnetic (mechanical) energy expenditure, which is why it is measured in units of Joules per second or foot-pound force per hour.

If we apply this argument to kinetic energy, we get:

$$P = \tfrac{1}{2}.m.v^2/t = \tfrac{1}{2}.m.R^2/t^3 = \tfrac{1}{2}.m \,.\, R/t \,.\, R/t^2 = \tfrac{1}{2}.m.v.a$$

If we apply this argument to angular (rotational) energy, we get:

$$P = 2\pi N.T = 2\pi/t \,.\, F.R = 2\pi \,.\, m.R/t^2 \,.\, R/t = 2\pi \,.\, m.R/t \,.\, R/t^2 = 2\pi.m.v.a$$

In both cases, power is momentum (M = m.v) multiplied by acceleration (a).

Electrical power is also the rate of energy expenditure, which is why it is measured in units of Joules per second or foot-pound force per hour. It may be calculated like this:

$$P = I.V$$

$$PE = P/f$$

$$F = PE/d$$

where:
'V' is the potential energy in a proton-electron pair (V = PE/e)
'I' is the electrical flow-rate inside a proton-electron pair (I = $e_v.f$)
'f' is the electron orbital frequency (f = v_N / $2\pi.R_N$)
'v' is the electron orbital velocity (v = $\sqrt{[T/X]}$)
'R' is the electron orbital radius (R = $X^R/\underline{T}$)
'g' is the potential acceleration within the proton-electron pair (g = $v_N{}^2/R_N$)

This means that for a given power rating, as electrical energy rises, its distance of influence must fall, which is the reason why electron orbital radius reduces with increasing electron velocity.

2.5.7 Field

Electrical and magnetic fields are radiated by particles

These formulas allow us to calculate the electro-magnetic field force between particles.

Originator	field force	@ 300K {N}
Henri Poincaré	$F = m_e.v^2/R$	6.73015473795726E-12
Isaac Newton	$F = G/\varphi \cdot m_e.m_p/R^2$	6.73015473795726E-12
Charles Augustin de Coulomb	$F = k.e^2/R^2$	6.73015473795726E-12
Joseph Henry	$F = \mu.I^2 \cdot (2\pi)^2$	6.73015473795726E-12
Hendrik Lorentz	$F = g.e/RC \qquad (g = v^2/R)$	6.73015473795726E-12

Each formula verifies the others because they all give exactly the same results.

2.5.8 Electro-Magnetic (EME)

If a shell-1 (N=1) electron temperature (T_1) is known, the electron temperature (T_N) in any shell may be calculated thus:

$$T_N = T_1/N \qquad \{K\}$$

Johannes Rydberg gave us relationships for electron energy (R_γ) and orbital radius (R_∞) that actually works. For example; the electron temperature at the Rydberg radius (a_o):

$$T = X^R/a_o = 33192.4000063507 \ K$$

Wavelength (λ) is defined by the orbital frequency (f) of the electron transferring it:

$$\lambda_N = c/f \qquad \{m\}$$
$$\lambda_N = 2\pi R_n \cdot (T_n/T_N)^{1.5} \qquad \{m\}$$
$$\lambda_N = (R_N/R_r)^{1.5} / R_\infty \qquad \{m\}$$

where: R_N is the electron orbital radius
where: R_r is an unknown orbital radius; 8.40016460895157E-11 m

Frequency (f) is defined by the orbital period of the electron transferring it:

$$f = 1/t_N = c/\lambda_N = v / 2\pi . R_N = g / 2\pi . v_N \qquad \{H_z\}$$

where: t_N is electron orbital period ($t_N = 2\pi . R/v$)

Amplitude (A) is equal to the orbital radius of the electron radiating it:

$$A_N = R_N \qquad \{m\}$$

Energy (magnitude) may be determined using the modified version of Planck's constant (h') or the orbital velocity (v) of the electron transferring it (refer to chapter 6.1):

$$E_N = h'/A_N \qquad \{J\}$$

$$E_N = \tfrac{1}{2} . m_e . v_N^2 \qquad \{J\}$$

$$E_N = a_o . R_\gamma/A_N \qquad \{J\}$$

Which is also the same energy as the orbiting electron's kinetic energy; KE.
Energy (range) = $2 \times E_N$ = KE - -KE = 2.KE

The Mathematical Laws of Natural Science

2.5.9 Charge

This Table identifies the *heat* energies for the proton-electron pair.

		Magnetic Charge			
RAM	H"	1"		1.00727638277233""	g/mol
		electron		proton	
m		9.1093897E-28		1.6726216378300E-24	g
RAM	m.N$_A$	5.44321026449229E-04		0.9994556789736	g/mol
R$_a$	RAM.R$_i$	15156.3563034305		8.2544164727608	J/g/K
R	m.R$_a$	1.38065156E-23		1.38065156E-23	J/K
c$_t$	R$_a$	15156.3563034305		8.2544164727608	J/g/K
C$_t$	m.c$_t$	1.38065156E-23		1.38065156E-23	J/K
c$_v$	N$_t$.c$_t$	22734.5344551458		12.3816247091411	J/g/K
C$_v$	m.c$_v$	2.07097734E-23		2.07097734E-23	J/K
c$_p$	c$_t$.c$_v$	37890.8907585763		20.6360411819	J/g/K
C$_p$	m.c$_p$	3.4516289E-23		3.4516289E-23	J/K
R$_m$	RAM.c$_p$	20.6248085507839		20.647279930507	J/K/mol
	R$_m$.L$_a$(T)	14.732228288661		14.732228288661	J/K/mol
	R$_i$.L$_a$(N$_t$)	14.732228288661		14.732228288661	J/K/mol
E	C$_p$.T$_3$/N$_t$	1.0077356686171E-22		1.0077356686171E-22	J
		# proton-electron pair		## hydrogen (H)	

This Table identifies the *charge* energies for the proton-electron pair.

		Electrical Charge			
RAC	H"	95736.1995661631""		95736.1995661631""	C/mol
		electron		proton	
q	e	1.60217648753E-19		2.94183820093364E-16	C
RAC	e.N$_A$	95736.1995661631		175786132.980978	C/mol
R$_a$	k$_B$/e	8.61735002820123E-05		4.69315939796359E-08	J/C/K
R	e.R$_a$	1.38065156E-23		1.38065156E-23	J/K
q$_t$	R$_a$	8.61735002820123E-05		4.69315939796359E-08	J/C/K
Q$_t$	e.q$_t$	1.38065156E-23		1.38065156E-23	J/K
q$_v$	N$_t$.q$_t$	1.29260250423018E-04		7.03973909694538E-08	J/C/K
Q$_v$	e.q$_v$	2.07097734E-23		2.07097734E-23	J/K
q$_p$	q$_t$.q$_v$	2.15433750705031E-04		1.1732898494909E-07	J/C/K
Q$_p$	e.q$_p$	3.45162890E-23		3.45162890E-23	J/K
R$_e$	RAC.q$_p$	20.6248085507839		20.6248085507839	J/K/mol
	R$_e$.L$_a$(T)	14.732228288661		14.732228288661	J/K/mol
	R$_i$.L$_a$(N$_t$)	14.732228288661		14.732228288661	J/K/mol
E	Q$_p$.T$_3$/N$_t$	1.0077356686171E-22		1.0077356686171E-22	J
		# proton-electron pair		## The Farad	

Both of the above Tables show that heat energy and charge energy **capacities** are identical; uniting the two properties (heat and charge) in terms of energy.

2.5.10 Microstates

Microstates (N) are the energy states of atomic particles that are governed by the relationship: $c_p.L_n(\underline{T}).RAM = K_B.N_A.L_n(N_t)$ J/K/mol

N:	N_t	N_v	N_p
relationship	$\exp(c_p.L_n(\underline{T})/R_a)$	c_v/R_a	c_Q/R_a
monatomic	1.5	1.5	2.5
diatomic	2.5	2.5	3.5
≥ triatomic	3.5	3.0	4

2.5.11 EME Deflection

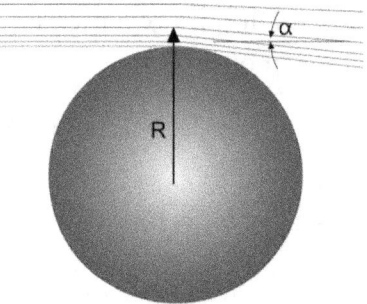

EME is emitted by a proton-electron pair in a direction normal to the electron's orbital plane according to the 'right-hand-rule', and travels in a straight line until deflected by magnetic charge, reflection or diffraction.

EME deflection by magnetic charge is calculated according to Newton's gravitational constant, and becomes significant as it passes massive celestial bodies; e.g. stars.

$$\alpha = \mathrm{Atan}(4 \cdot a_o/R \cdot m_s/m_u) \quad \{\mathrm{m/m} \cdot \mathrm{kg/kg}\}$$

where:
$G = a_o \cdot c^2/m_u$ $\{m^3 / kg.s^2\}$
m_s = the mass of our sun $\{kg\}$
m_u = unit mass of ultimate density $\{kg\}$
R = distance from the centre of mass $\{m\}$
a_o = Rydberg's radius $\{m\}$

Because EME radiated by the proton-electron pairs in the innermost atoms within a body gets bounced around by its neighbouring pairs, it will take much longer for core temperatures to stabilise. This is why the surface of a body cools faster than its centre. It is also why EME appears to slow down when passing through matter such as water or glass. In reality, EME does not slow down, it simply gets bounced around.

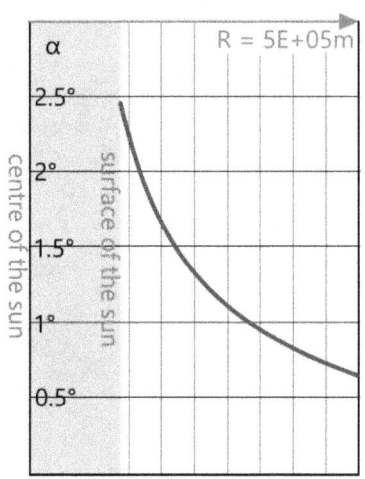

2.5.12 Temperature

The following Table lists the key universal temperatures:

	Formula:	Temperature (K)	electron velocity (m/s)	orbital radius (m)
		$T = PE/k_B'$	$v = \sqrt{[T/X]}$	$R = X^R/T$
abs. minimum	$T_x = X.(c / Y.\xi_m)^2$	2.0427490756827	17162.242521927	8.5985409857E-07
e' = e	$T_e = T_n.(e / m_p.RC)^2$	339468.852842837	6996268.84651132	5.17416001606939E-12
Rydberg radius	$T_e = X^R/a_o$ $a_o = R_n.(\xi_v/4\pi)^2$	33192.400006351	2187690.3505355	5.2917721067E-11
Planck min.	$T_o = X.v_o^2$ $v_o = v_n/\xi_v$	210.19332853584	174090.86662108	8.3564315638E-09
Planck mean	$T_m = X.v_m^2$ $v_m = v_n / \sqrt{.\xi_v}$	361962.55467156	7224342.80705	4.8526184336E-12
Neutronic	$T_n = m.c^2/k_B'$ $v_n = c$	623316124.71718	299792459	2.81793795384E-15

Absolute Minimum: the temperature at which a proton can no longer hold onto its orbiting electron partner.

e' = e: electron-electron attraction is equal to proton-proton repulsion

Rydberg Radius: the temperature that gave us the dynamic ratio; ξ_v.

Planck Minimum: the gas-transition temperature of the largest noble atom.

Planck Mean: the mean value between his minimum and the neutronic.

Neutronic: the temperature at which a proton-electron pair will unite as a neutron.

Given the following:

EME velocity = 299792459 m/s (100% x c)
'Big-Bang' initial velocity ≈ 1773498 m/s (0.6% x c)
velocity today of all universal matter ≈ 230000 m/s (0.08% x c)

The EME radiated at the last 'Big-Bang' has today travelled 1.30274E+26m, which is well beyond the limits of the universe today (4.353083E+23m). The temperature measured during the Cobe project (≈2.7255 K) had nothing to do with heat left over from the last 'Big-Bang', it was simply detecting the EME (heat) radiated by all the stars and planets in the universe.

The Mathematical Laws of Natural Science

2.5.13 $E=mc^2$

Henri Poincaré's '$E=mc^2$' applies to the creation of neutrons.

Henri Poincaré derived his formula at the end of the 19th century, but he wrote it like this;

$c = \sqrt{[E/m]}$ {m/s}

However, he was unsure of its meaning.

The difference between kinetic and dynamic energies looks like this:

$KE = \frac{1}{2}.m.v^2$ {J}
$PE = m.g.d$ {J}

According to Newton's laws of orbital motion, the potential acceleration between a satellite and its force-centre in elliptical orbits looks like this:

$g = v^A.v / d.(1-e)$ {m/s²}   (chapter 9.8.3)

But in circular orbits:

$v^A = v$ & $e = 0$ and $g = v^2/d$ {m/s²}

Therefore, the potential energy between a force-centre and its satellite in circular orbits is:

$PE = m.(v^2/d).d$ **{J}**

i.e.:

$PE = m.v^2$ & $KE = \frac{1}{2}.m.v^2$
$PE = 2.KE$

Therefore, Poincaré's formula applies to circular orbits, and works like this:

$E = PE = 2.KE = 2 . \frac{1}{2}.m_e.v^2 = m_e.v^2$ {J}

... and when the orbiting satellite achieves 'light-speed' (c), such as for proton-electron pairs in the core of bright stars and planets, this formula becomes:

$E = m_e.c^2$ {J}

at which instant, the proton-electron pair unites to create a neutron. This is the meaning of Henri Poincaré's formula.

Energy and mass do not transpose with velocity.

3 Electricity

Electricity is the transfer of electrical energy, that may be in the form of EME transmission (AC) or electron transfer (DC). Electricity always travels from negative to positive.

In both cases, because EME possesses electrical *and magnetic* energy, and electrons possess both electrical *and magnetic* charge, all forms of electrical energy are always accompanied by magnetic energy.

All forms of electricity are polar; like poles repel and opposite poles attract. This attraction is evident in the proton-electron pair, where the proton possesses positive charge and the electron possesses negative charge.

It is important to remember that when pulling electrons through a conductor (DC), their electrical (e) and magnetic (m) charges flow together. And similarly, the EME radiated by AC electricity is 'electro-magnetic energy'.

Today, we allocate their units of measurement like this:

electricity		
label	symbol	units
current	I	C/s
electro-motive force	emf	J/C
resistance	Ω or R	$J.s/C^2$

The Mathematical Laws of Natural Science

3.1 Electrical Charge

Electrical charge is shared between Quanta.

Every electron possesses a negative electrical charge (e^-) of exactly the same magnitude, everywhere in the universe.

Every lone proton possesses a positive electrical charge (e^+) of exactly the same magnitude, everywhere in the universe.

Electrical charges attract or repulse all other electrical charges throughout the universe, dependent upon polarities.
  Same polarity charges repel.
  Opposite polarity charges attract.

The proton's additional magnetic charge gives it the capacity to collect additional electrical charge from the EME transferred by its electron partner. The proton uses this additional electrical charge to repel neighbouring atomic protons (gas).

When it loses its electron partner, a proton's electrical charge (magnitude) will fall to that of the electron; the elementary charge unit (e^+).

The metric and Imperial unit for the measurement of electrical charge is the Coulomb.

Because electrical charge is shared between Quanta, the potential energy between each of a million Quantum neighbours is one-millionth that of two neighbouring Quanta. Electrical charge sharing is the reason we see negligible electrical interaction between celestial bodies.

The electrical **charge** (e') in a proton when partnered with an orbiting electron varies linearly with the kinetic energy of its electron partner:

$$e' = m_p.RC \cdot \underline{T/T_n} \qquad \{C\}$$

which maximises at: $e_n = m_p.RC \quad \{C\}$

A proton's electrical charge varies between; $e^+ \leq e' \leq e_n^+$ whilst partnering an orbiting electron,

e_n is the maximum electrical charge a proton can hold, at which its orbiting electron partner will have achieved light-speed (c) uniting both particles as a neutron.

3.2 Direct Current (DC)

DC electricity is the flow of electrons. It occurs naturally within an atom's proton-electron pairs, and unnaturally along an electrical conductor.

The difference between natural and unnatural electricity is; natural DC is contained and present in all atoms at all times, whereas unnatural DC is transferred between adjacent atoms by an external potential energy (or difference).

Natural DC electricity within a proton-electron pair is simply the orbital frequency of an electron, it is defined and calculated in the same way you would for unnatural DC. Electron flow rate within a proton-electron pair is based upon the proton-electron pair's temperature ($\underline{T}$), which - within an atom - is dependent upon its shell number ($\underline{T}_N = \underline{T}/N$):

Unnatural DC electricity within a conductor is the rate of flow of electrons between adjacent atoms. It is defined by a potential energy across its terminals. The potential energy (or difference) at each end of the conductor must be of equal magnitude and opposite polarity. I.e. you must allow the electrons to travel around a circuit. You cannot simply keep on pulling electrons from the conductor's atoms unless you push the same number of electrons back in at the other end.

Because an atom's outermost shell electrons have the lowest potential energy, these are the electrons that will be released from the conductor's atoms. The potential energy applied at the positive terminal will pull these electrons from all the atoms across the conductor's cross-sectional area. Each resultant lone nucleic proton in these atoms will pull an electron from its adjacent atom, making it positively charged, etc. all along the conductor towards the negatively charged terminal.

Due to electrical sharing, if the magnitude of the applied potential energy is insufficient to pull an electron from all of the outermost atomic shells across the conductor's cross-sectional area, electrical current will not flow. For example; you must apply more than 0.0664708917 Volts across a 4.6E-05m diameter tungsten filament at 300K in order to generate electron flow.

The Mathematical Laws of Natural Science

As you increase the potential energy (Voltage) across a conductor, the electron transfer velocity will increase accordingly; $v = \sqrt{[2.PE/m_e]}$ per electron. But, if the electron transfer velocity is faster than the orbital velocity of the receiving atom's outermost shell at its temperature, it will slow down once in orbit about its new proton partner, releasing its kinetic energy in the form of electro-magnetic energy (EME), which will be absorbed by all the other atomic shell electrons until electrical forces balance. It is this release of EME that increases the temperature of a conductor with increasing Voltage.

Have you ever wondered why a tungsten filament looks white, when its *temperature* is 2,884K, at which only two proton-electron pairs per atom orbit at 1.685087E+14 Hz, emitting EME in the infra-red range. All the other proton-electron pairs emit darker (colder) EME.

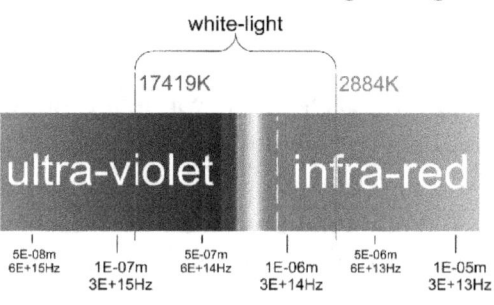

Let's take a look at a 120V, 60W filament:
The total atomic Voltage at 2,884K is less than 20V, so what happens to the remaining supply Voltage (100V)?

The documented *temperature* of 2,884K, only applies to its surface; that which we measure. Its core temperature is 6 times higher (6 x 20 = 120), giving it a core temperature of 17,419.11136K, the frequency of which is 2.501416E+15 Hz; emitting ultra-violet EME. This filament therefore emits EME across the entire visible spectrum, which is the reason why its radiated light is white.

Yes! the core atoms are theoretically gaseous at 17,419.11136K, but they are contained in a viscous state within the viscous outer matter that together comprise more than 90% of the material at less than 6,200K. This is why the neutron energy cell will work; even if you heat up the core atoms to a gaseous state, most of the fuel will remain viscous; just as with a tungsten filament.

3.3 Alternating Current (AC)

Every proton-electron pair naturally and continually, generates and radiates, electro-magnetic energy. If the shell-1 proton-electron pairs of an electrical conductor are aligned, they will act together, unidirectionally, significantly increasing their combined directional strength.

In AC circuits, electrons do not flow along a conductor; EME travels along the conductor in the direction of shell-1 proton-electron pair alignment according to the right-hand rule. Magnetism aligns the proton-electron pairs and the direction of travel defines the polarity in the conductor. Because EME is electro-magnetic energy, electricity and magnetism will both exist coincidently along the length of the conductor.

As the potential difference (+ve : -ve) across a conductor rises, the kinetic energy of its orbital electrons will also rise, increasing the temperature of the conductor's proton-electron pairs, and with it; the magnitude of the radiated EME. In AC circuits, current (I) is defined by the orbital velocity of the electrons in the conductor's proton-electron pairs, and Voltage (V) is defined by the potential difference across the circuit. And because electrical resistance is defined thus; $R = V/I$; electrical resistance also increases with temperature in AC circuits.

Moreover, whilst DC energy is defined by the extraction energy in an atom's outermost electrons, i.e. those with the lowest potential energy, the energy in AC electrical circuits is defined by the EME radiated by an atom's innermost (shell-1) proton-electron pairs. This is why AC electrical current is stronger than DC current for a given Voltage; AC electrical resistance is lower.

The Mathematical Laws of Natural Science

3.4 Calculations

The properties that define electrical performance through elemental matter, are power, Voltage and resistivity; current and resistance are consequences. Their relationships are:

power = V.I		{J/s}
Voltage: V = I.Ω		{J/C}
resistivity: ρ = Ω.A/ℓ		{J.s.m/C²}
current: I = V/Ω		{C/s}
resistance: Ω = V/I		{J.s/C²}

Apart from resistivity, all the above are known (exactly) for; the proton-electron pair, the atom and elemental matter.

The electrons involved in DC electricity are those in an atom's outermost shell (proton-electron pairs).

Voltage within a proton-electron pair is the potential energy between the orbiting electron and its proton partner;

Voltage across a conductor is the potential energy required to extract an outermost electron from an atom's shell ('N' below refers to outermost shell number);

$$V = PE/e = m_e.v^2/e = m_e.\underline{T}_N / X.e \;\{J/C\}$$

Current is the orbital frequency of the electron partner;
$$I = e.f = e \cdot (\underline{T}_N/\underline{T}_n)^{1.5} / t_n \;\{C/s\}$$

Resistance is the energy preventing electron release;
$$\Omega = V/I \;\{J.s/C^2\}$$

Power is the frequency of potential energy in the proton-electron pair;
$$P = V.I = PE.f \;\{J/s\}$$

Resistivity applies to a particular element based upon its atomic spacing;
$$\rho = \Omega \cdot d^2/2\pi R \;\{J.s.m/C^2\}$$

The Mathematical Laws of Natural Science

3.4.1 Resistivity

Electrical *resistivity* (ρ) only applies to the transfer of an atom's outermost electrons between adjacent atoms due to an external potential energy (or difference).

In an electrical conductor, resistivity is calculated thus:

$\rho = \Omega.A/\ell$ {J.s.m/C^2}

where: Ω is the conductor's electrical resistance, A is its cross-sectional area and ℓ is its length

The inter-atomic equivalent of which looks like this:

$\rho = \Omega_a \cdot (d^2 / 2\pi R)$ {J.s.m/C^2}

where: Ω_a is the atom's shell-N electrical resistance, R is the atom's shell-N electron's orbital radius and d is the average inter-atomic distance ($\sqrt[3]{[m_a/\rho]}$).

The average disparity between the above formula and documented values for all 92 elements is ≈ 1.1

Resistivity	copper	silver	gold	iron	tungsten	aluminium
documented	1.70E-08	1.63E-08	2.44E-08	9.70E-08	5.65E-08	2.70E-08
calculated	9.462E-08	9.578E-08	7.384E-08	1.016E-07	7.329E-08	1.741E-07
disparity	5.566	3.026	3.172	1.048	1.297	6.448
Resistivity Table (@273.15 K): $\rho_e = \Omega \cdot d^2 / 2\pi R$ J.s.m/C^2						

The disparity in the above Table is due to the following:
1) Documented values are subject to experimental error (see next page).
2) Documented values are generally copies of other documented values.
3) Calculated values are based upon pure crystals of the base metal.
4) Calculated values are based upon undeformed base metal.
5) Calculated values are based upon average (directional) atomic distances.
6) Calculated values are based upon an operating temperature of 300K

Page 43

The Mathematical Laws of Natural Science

This image, which was issued for copper by CINDAS (Centre for Information and Numerical Data Analysis and Synthesis), reveals how experimental error can vary documented values by as much as 1E±10.

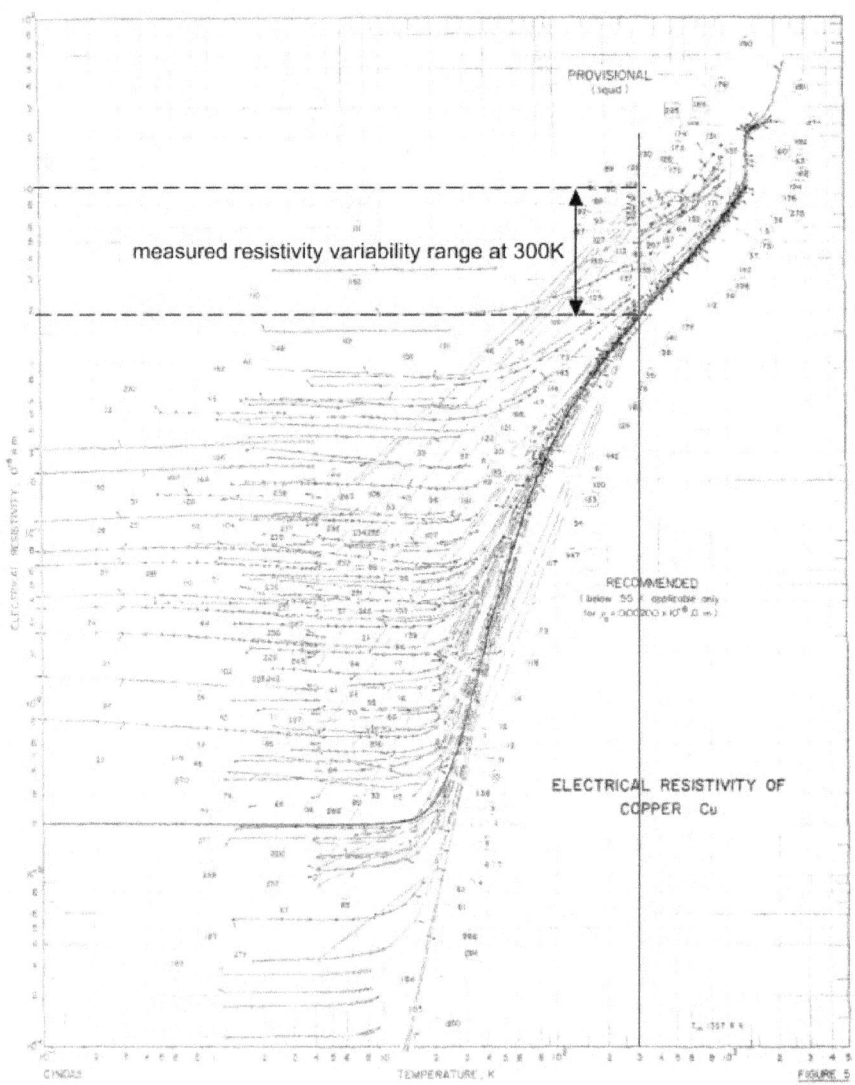

Whereas the disparity between the minimum documented value for the electrical resistivity of copper (1.7E-08 J.s.m/C²) and the calculated value is just 1.56. The important consideration here is that the metal concerned is *copper*, the most common electrical conductor in the world; which is subject to world-wide measurement techniques that can vary significantly.

4 Magnetism

Magnetism is the transfer of magnetic energy, that may be in the form of EME transmission (AC) or electron transfer (DC). Magnetism always travels from positive to negative.

In both cases, because EME possesses magnetic *and electrical* energy, and electrons possess both magnetic *and electrical* charge, all forms of magnetic energy are always accompanied by electrical energy.

Magnetic charge is non-polar, and only attracts other magnetically charged particles. Magnetic field is polar, like poles repel and opposite poles attract.

Magnetic field is generated by proton-electron pairs. Accumulation and alignment of shell-1 atomic proton-electron pairs will increase field strength, but decrease its field of influence.

Electrical and magnetic equivalent terminologies are today explained thus:
   magnetic flux (Φ) ≡ electrical current (I);
   magneto-motive force (mmf) ≡ electro-motive force (emf);
   magnetic reluctance (R) ≡ electrical resistance (R).

However, there is a problem with today's understanding of magnetism:

It is important to remember that when pulling electrons through a conductor (DC), their electrical (e) and magnetic (m) charges flow together. And similarly, the EME radiated by AC electricity is 'electro-magnetic energy'.

This is why electricity and magnetism always occur together. Their units of measurement must therefore be direct equivalents, taking their respective properties into account. But this is not the case today.

Today, we allocate their units of measurement like this, whereas they should look like {this}:

magnetism				electricity		
label	symbol	units		label	symbol	units
flux	Φ	J.s/C	{kg/s}	current	I	C/s
magneto-motive force	mmf	C/s	{J/kg}	electro-motive force	emf	J/C
reluctance	R	$C^2/J.s^2$	{J.s/kg^2}	resistance	Ω or R	J.s/C^2

Where the unit 'kg' is used for magnetic charge; *mass is magnetic charge.*

4.1 Magnetic Charge

Magnetic charge is accrued between Quanta.

Every electron possesses a non-polar magnetic charge (m_e) of exactly the same magnitude, everywhere in the universe.

Every proton possesses a non-polar magnetic charge (m_p) of exactly the same magnitude, everywhere in the universe.

Magnetic charges attract all other magnetic charges throughout the universe. Non-polar charges only attract. We currently refer to this attraction as gravity.

Their relative charge magnitudes are; $m_p = \xi_m . m_e$

The metric unit for the measurement of magnetic charge is the kilogram, and the Imperial equivalent is the pound.

Because magnetic energy is accrued between Quanta, the magnetic attraction between one million Quanta is one million times stronger than between any two other quanta. Magnetic charge accrual is the reason for such strong attraction between celestial bodies.

Where atomic shell-1 proton-electron pairs can be aligned within a body due to their atomic lattice structures, they will produce a strong magnetic field. Such alignment in iron (with a BCC lattice structure) must be induced artificially, e.g. with a lodestone, but it occurs naturally in metals with an HCP structure such as Cobalt, Dysprosium, Gadolinium, Neodymium and Terbium. This is the magnetism that exists in bar-magnets. Whilst this form of magnetism is much stronger than magnetic charge, it has a very limited field of influence, which reduces with increasing strength.

The influence of magnetic charge, however, is ever-present and constant throughout the universe (via the great attractor), gradually pulling together all its matter, i.e. slowing down the expansion caused by the last 'Big Bang'. Eventually, all the celestial bodies in the universe will stop moving away from each other and gradually re-accrete into another ultimate-body. A new universal period will then begin. No external assistance is required for this repetition to continue indefinitely.

4.2 Mass

Non-polar magnetic charge is what we today refer to as mass.

The term 'mass' is an unknown concept. It is today used to describe the inertia of matter; i.e. its resistance to movement. This resistance is actually the reciprocity between a body's magnetic charge and the magnetic field in its universal environment.

I.e. mass (m) is magnetic charge (m), the magnitude of which is the non-polar equivalent of the elementary charge unit: $m \equiv |e|$

Therefore; the elementary charge unit (±e) should be referred to as the electrical charge unit (e), and mass should be referred to as the magnetic charge unit (m). The non-polar nature of the magnetic charge is what generates attraction between all particles.

Because the magnetic charge in Quanta is constant, the magnetic field they generate will also be constant; just as mass is constant. And we currently refer to this magnetic field as gravity.

In order to harmonise magnetic and electrical charges, we could allocate a new unit of measurement to the magnetic charge, e.g. the 'Gilbert', giving it the same numerical value as the elementary electrical charge unit; 'e' (Coulomb):

$$\gamma = e/m_e = 1.75881869180545E+11 \text{ G/kg}$$

Where 'γ' is the factor used to numerically convert mass to the magnetic charge (m). I have chosen the unit-name 'Gilbert', with the symbol; 'G', in deference to William Gilbert. I.e.:

$$m_e = 1.60217648753E-19 \text{ G and } 1kg = 1.75881869180545E+11 \text{ G}$$

Isaac Newton's magnetic constant is currently calculated like this:

$$G = a_o.c^2/m_u = 6.67359232004333E-11 \text{ m}^3 / \text{kg.s}^2$$

and his gravitational force is calculated thus:

$$F = G.m_1.m_2/d^2 \qquad \{kg.m/s^2\}$$

The Mathematical Laws of Natural Science

But we could revise Newton's constant and the magnetic charge in terms of Gilberts thus:

$$M = a_o.c^2 / m_u.\gamma^2 = 2.15733469430661\text{E-}33 \quad m^3/G.s^2$$

$$m = m.\gamma \qquad \{G\}$$

Newton's (and Gilbert's) force formula still applies, but; m_1 & m_2 are now the magnetic charges of each body, measured in Gilberts.

Example:

We currently calculate the **Coupling Ratio** (φ) using mass & 'G' as follows;

$$\varphi = G.m_e.m_p / k.e^2 = \mathbf{4.40742111792334\text{E-}40}$$

But it could be calculated using magnetic charge & 'M';

$$m_e = 1.60217648753\text{E-}19 \; G$$

$$m_p = \xi_m \times 1.60217648753\text{E-}19 \; G$$

$$\varphi = M.m_e.\xi_m.m_e / k.e^2 = M.m_e^2.\xi_m / k.e^2$$

because m_e & e have the same numerical values (1.60217648753E-19)

$$\varphi = \xi_m.M/k = \mathbf{4.40742111792334\text{E-}40}$$

4.2.1 Constant @ Any Temperature

The magnetic force in a proton-electron pair is defined by Joseph Henry:

$$F_m = \mu.I^2.(2\pi)^2 \qquad \{kg.m/s^2\}$$

$$\mu.I^2 = (m_e.R/e^2) \times (e.2\pi f)^2 \times (4\pi.R^2 / 4\pi.d^2) \qquad \{kg.m/s^2\}$$

remove the constants at any distance 'd' and we get:

$$\text{factor} = R.f^2.R^2 = R^3.f^2 = 6.41524280848628 \; m^3/s^2$$

which is the reciprocal of Isaac Newton's constant of proportionality for the proton-electron pair:

$$K = t_n^2/R_n^3 = 1/\text{factor} = 0.15587874533403 \; s^2/m^3$$

i.e. a constant!

The Mathematical Laws of Natural Science

4.3 Gravity

Gravity is the magnetic field of attraction radiated by magnetic charge.

Let's see if we can demonstrate that gravity is magnetism by comparing magnetic field force as described by Joseph Henry with gravitational force as described by Newton.

Gilbert & Newton: $\quad F_N = G \cdot m_1 \cdot m_2 / d^2 \qquad \{kg \cdot m/s^2\}$
Henry's magnetic field: $\quad \mu_o = m_e \cdot R_n / e^2 \qquad \{kg \cdot m/C^2\}$
Magnetic field in nucleic protons: $\mu_p = m_p \cdot d / e^2 \qquad \{kg \cdot m/C^2\}$
Where; m_p is the mass of a proton and $d = R_1$ = orbital radius of shell-1 electrons

Newton's potential force: $\quad F_N = G/\varphi \cdot (m_p/d)^2$
Henry's potential force: $\quad F_H = \mu_p \cdot I^2 \cdot (2\pi)^2$
where:

Newton's Force (F_N)	Henry's Force (F_H)
$d = X^R / T \quad \{m\}$	$I = e \cdot f \quad \{C/s\}$
$G = a_o \cdot c^2 / m_u \quad \{m^3/kg \cdot s^2\}$	$f = v/(2\pi d) \quad \{m/s^2\}$
$\varphi = 4.40742111792334E\text{-}40$	$v = \sqrt{[T/X]} \quad \{m/s\}$
$m_p = 1.67262163783E\text{-}27 \; kg$	T_1 = Shell-1 temperature $\{K\}$

Both of which give the following calculation results:

@ T (K)	F_N (kg.m/s²)	F_H (kg.m/s²)
T_x	5.7295566068112E-13	5.7295566068112E-13
300	1.23575813653591E-08	1.23575813653591E-08
5788	4.59989601440376E-06	4.59989601440376E-06
T_n	5.33467164189259E+04	5.33467164189259E+04

from which, we can conclude that gravity is indeed magnetism.

The Mathematical Laws of Natural Science

4.4 Calculations

Joseph Henry's magnetic constant (μ_o) gives us the minimum magnetic field; at the neutronic condition:

$$\mu_o = m_e \cdot R_n / e^2 \qquad \{kg.m/C^2\}$$

<small>where; m_e = electron mass & e = electron charge & R_n = neutronic radius</small>

and magnetic force within **a proton-electron pair** at any temperature is:

$$F_m = \mu \cdot I^2 \cdot (2\pi)^2 = m_e \cdot v^2 / R \qquad \{kg.m/C^2 \times (C/s)^2 = kg.m/s^2\}$$

<small>Where; $\mu = m_e \cdot R/e^2$ & $I = e.f$</small>

Between adjacent atoms:

$$F_m = \mu_o \cdot \xi_m / \zeta^3 \cdot I_s^2 \cdot (2\pi)^2 \cdot (R/d)^3 \cdot RAM \cdot m_n / m_p \qquad \{kg.m/s^2\}$$

<small>Where; d is the mean inter-atomic spacing ($\sqrt[3]{[m_a/\rho]}$) & R is the electron orbital radius</small>

which gives exactly the same result as;

$$F_m = \rho \cdot h_e^2 / \zeta^3 \qquad \{kg.m/s^2\}$$

<small>Where; 'h_e' is Isaac Newton's constant of motion for an orbiting electron</small>

4.4.1 Bar Magnets

The force exerted by a bar magnet on a ferro-magnetic target may be calculated like this:

$$F_m = m_p \cdot R_1 \cdot f_1^2 \cdot (2\pi)^2 \cdot N° \cdot (R_1/d)^2 \qquad \{kg.m/s^2\}$$

<small>Where;
proton mass: m_p
shell-1 electron orbital radius: R_1 ($R_1 = X^R \cdot T$)
shell-1 electron orbital frequency: f_1 ($f_1 = [T/T_n]^{1.5} / t_n$)
number of atoms in the bar magnet: $N°$ ($N° = m/m_a$)
distance between bar magnet and target: d</small>

4.4.2 Solenoids and Coils

The force exerted by a solenoid plunger may be calculated like this:

$$F = \mu_1 \cdot I^2 \cdot (2\pi)^2 \qquad \{kg.m/s^2\}$$

<small>Where:
Henry's magnetic field: μ_1 ($\mu_1 = m_e \cdot R_1/e$)
shell-1 electron orbital radius: R_1 ($R_1 = X^R \cdot T$)
I = applied current</small>

5 Atomic Particles

The entire universe comprises just two quanta; the proton and the electron.

All quanta are homogeneous packets of electro-magnetic charge.

Quanta shall be the collective term used for atomic particles.

Only two Quanta are required to make our universe work perfectly, exactly as we see and feel it; the proton and the electron.

All Quanta are packets of electro-magnetic charge.

Quanta are not made up of smaller sub-atomic particles.

The difference between the strength of attractive and repulsive forces due to magnetic and electrical charges is the coupling ratio (φ).

Lone Quanta cannot emit or collect electro-magnetic energy (EME).

The atom comprises proton-electron pairs that between them hold it together and naturally attract or repel other atoms.

Quanta are pictorially represented here as solid spherical objects for convenience only. It is not proposed here that they are in any way spherical or solid.

The Mathematical Laws of Natural Science

5.1 Electron

All electrons are identical.

An electron is a packet of electro-magnetic charge.

All electrons are identical.

An electrons magnetic charge is constant and non-polar.

An electrons electrical charge is constant and negative.

Electrons only exist in free-flight or as part of a proton-electron pair.

In free-flight, electrons travel at the velocity they had achieved at the moment they escaped their proton-electron partnership. Their kinetic energy will not change until they again become partnered with a proton.

When partnered with a proton, electrons convert the energy they collect from the EME in their surroundings, into kinetic energy.

Electrons do not exist in outer space. Their magnetic charge ensures that they are always attracted to the nearest celestial body.

The electron's properties are listed below.

Mass: m_e = 9.1093897E-31 kg

Electrical charge: e = -1.60217648753E-19 C

Density: ρ = 7.1266079635045E+16 kg/m³

Radius: r = 1.45046059426276E-16 m

Volume: V = 1.27822236702922E-47 m³

Polar moment of inertia: J = 7.665864560566651E-63 kg.m²

Lone electrons cannot be coalesced into viscous matter because they all possess a negative electrical charge. They can only exist as a gas.

The Mathematical Laws of Natural Science

5.1.1 Photon

There are no such things as photons.

A photon is said to be an electron travelling at the speed of light, and is the source of all light.

Because an electron has a fixed mass, and light travels at a constant velocity, its kinetic energy would have to be constant:

$$KE = \tfrac{1}{2}.m_e.c^2$$

If this were true; the electro-magnetic spectrum would not exist.

Moreover, if light is indeed photons, our sun would have exhausted all of its electron mass in the first seconds of its life as a bright star. And all of the neutrons and remaining protons would have reverted to lone particles. The sun would no longer be able to hold onto its satellites. The solar system would not exist.

For example:

Our sun radiates;

E_s = 9.9252916097005E+33 Joules of energy.

which equates to a power of;

$P = E_s.f$ = 4.75535480329058E+48 J/s.

Our sun has;

$N° = m_s / (1+ \phi_{iron}).(m_p+m_e)$ ≈5.53425772798E+56 electrons.

The kinetic energy of an electron @ our sun's surface temperature (5788 K);

$KE = \tfrac{1}{2}.m_e.v^2 = \tfrac{1}{2}.m_e.\underline{T}/X$ = 3.80120117845959E-19 J

based on which, all of its electrons represent a total energy of;

$E = KE.N°$ = 2.10368269974967E+38 J

it would have lost all of its electrons;

t = E/P = 4.42381859350217E-11 seconds.

Alternatively, the KE of an electron @ the neutronic temperature ($\underline{T}_n$);

$KE_n = \tfrac{1}{2}.m_e.c^2$ = 4.09355561131267E-14 J

based on which, all of its electrons represent a total energy of;

$E = KE_n.N°$ = 2.2654791776823E+43 J

it would have lost all of its electrons in just;

t = E/P = 4.76405919515123E-06 seconds.

5.2 Proton

All protons are identical.

A proton is a packet of electro-magnetic charge.

All protons are identical.

A proton's magnetic charge is constant and non-polar.

A proton's electrical charge is positive.

A proton's electrical charge is constant and equal to that of an electron whilst it is not partnered with an electron.

When partnered with an electron, a proton's electrical charge varies with the kinetic energy in its orbiting electron.

The additional electrical charge collected from its electron partner is held and used to repel adjacent protons (atomic nuclei).

Protons have no kinetic energy of their own. They move only when caused to do so by external means.

Protons do not exist in outer space. Their magnetic charge ensures that they are always attracted to the nearest celestial body.

The proton's properties are listed below.

Mass: m_p = 1.67262163783E-27 kg

Electrical charge (alone): e = 1.60217648753E-19 C

Electrical charge (with electron partner): e' = m_p.RC . $\sqrt{[T/T_n]}$ {C}

Electrical charge (maximum): e_n = m_p.RC = 2.94183820093364E-16 C

Density: ρ = 7.1266079635045E+16 kg/m^3

Radius: r = 1.77613270336827E-15 m

Volume: V = 2.34700946985653E-44 m^3

Polar moment of inertia: J = 2.11061258698748E-57 kg.m^2

Lone protons cannot be coalesced into viscous matter because they all possess a positive electrical charge. They can only exist as a gas.

Lone protons cannot hold onto a neutron.

The Mathematical Laws of Natural Science

5.2.1 Hydrogen Atom

Lone protons are frequently referred to as hydrogen gas atoms. They are no such thing; a lone proton is a lone proton. 99.97% of what we today refer to as hydrogen gas comprises lone protons.

Lone protons can only exist as a gas due to their positive electrical charges.

5.2.2 Hypothesis

It is perhaps possible, that a proton comprises two particles; a positively [electric] charged positron of 9.1093897E-31 kg and an outer body of 1.67171069886E-27 kg (magnetic charge) that possesses no electrical charge. If so, it is this non-charged outer-body that collects additional electrical charge from its orbiting electron partner and repels neighbouring protons.

5.3 Neutron

A neutron is a proton-electron pair united through high temperature.

A neutron is created, and can only exist within an atom.

When split, a neutron will revert to its particles releasing its stored energy.

A neutron is a proton-electron pair united through high temperature (T_n).
All neutrons are identical.
The electrical charge in a neutron is zero.
The magnetic charge in a neutron is constant and non-polar.
Neutrons have no kinetic energy of their own.
Neutrons are only created, and can only exist, within an atom.
Neutrons store all the kinetic, potential and spin energy their component parts possessed at the time of their creation.

The neutron's properties are listed below.

Mass: m_n = 1.6735325768E-27 kg

Electrical charge: e = 0 C

Density: ρ = 7.1266079635045E+16 kg/m³

Radius: r = 1.77645508248591E-15 m

Volume: V = 2.34828769222356E-44 m³

Polar moment of inertia: J = 2.11252872891479E-57 kg.m²

Stored energy: E = 1.63785606465701E-13 J

Neutrons are not particles in their own right. They cannot be picked up (trapped) by accident or design. They are [only] created inside an existing atom, and they cannot exist outside one. When ejected from their nucleus, they will revert to their original component parts, a proton (alpha-particle) and an electron (beta-particle), or they will revert to a proton-electron pair if trapped within their parent atom.

5.3.1 Neutron Creation

Neutrons are created when magnetic field force exceeds centrifugal force.

The electrical charge between a proton and its orbiting electron generates a potential force (F_e) that follows Coulomb's law regardless of orbital radius. The magnetic potential force (F_m), however, varies with orbital radius:

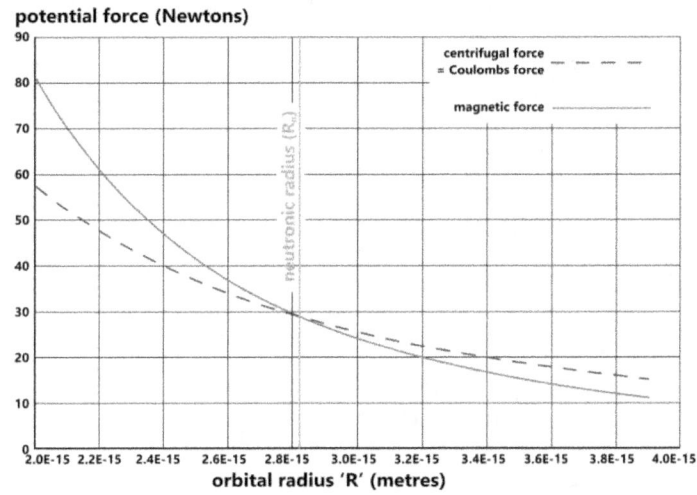

At orbital radii greater than the neutronic radius (R_n), the balancing potential forces of an orbiting electron are as follows: $F_G = F_e = F_c > F_m$

At the neutronic radius (R_n), the balancing potential forces of an orbiting electron are as follows: $F_G = F_e = F_c = F_m$

At orbital radii less than the neutronic radius (R_n), the balancing potential forces of an orbiting electron are as follows: $F_G = F_e = F_c < F_m$

Where the forces are calculated as follows:
Potential: $F_G = -G.m_e.m_p / \varphi.R^2$ $\{m^3/kg.s^2 . kg^2/m^2 = kg.m/s^2 = N\}$
Electrical: $F_e = -k.e^2 / R^2$ $\{kg.m^3/C^2.s^2 . C^2/m^2 = kg.m/s^2 = N\}$
Centrifugal: $F_c = m_e.v^2 / R$ $\{kg.m^2/s^2 / m = kg.m/s^2 = N\}$
Magnetic: $F_m = \mu.g.e^2 / R$ $\{kg.m/C^2 . m/s^2 . C^2/m = kg.m/s^2 = N\}$
        $= m_e.c^2.R_n^2 / R^3$ $\{kg.m^2/s^2 . m^2/m^3 = kg.m/s^2 = N\}$
and:
     $g = -G.m_p / R^2.\varphi$ $\{m^3/kg.s^2 . kg/m^2 = m/s^2\}$
     $\mu_o = R_n.m_e/e^2 = 1E\text{-}07 \text{ kg.m}/C^2$

The Mathematical Laws of Natural Science

where 'R' is the electron's orbital radius.

5.3.2 Neutron Energy

A neutron stores the energy it was generating at the time of its creation.

Every neutron holds 1.637856064657O1E-13 Joules of energy.

The energy stored in all neutrons may be calculated as follows:

$$E_n = |KE| + |PE| + |SE|$$

Kinetic Energy:

$$\begin{aligned} KE &= \tfrac{1}{2} \cdot m_e \cdot c^2 \\ &= \tfrac{1}{2} \times 9.1093897E\text{-}31 \times 299792459 \\ &= 4.09355561131267E\text{-}14 \text{ J} \end{aligned}$$

Potential Energy:

$$\begin{aligned} PE &= -m_e \cdot c^2 \\ &= -9.1093897E\text{-}31 \times 299792459 \\ &= -8.18711122262534E\text{-}14 \text{ J} \end{aligned}$$

Proton Spin Energy:

$$\begin{aligned} SE_p = E_3 &= KE_e + PE \\ &= 4.09355561131267E\text{-}14 \text{ J} + -8.18711122262534E\text{-}14 \\ &= -4.09355561131267E\text{-}14 \text{ J} \end{aligned}$$

Electron Spin Energy:

$$SE_e = E_0 - E_1 - E_3$$
$$J_e = \tfrac{2}{5} \cdot m_e \cdot r_e^2 = 7.66586456056651E\text{-}63 \text{ kg.m}^2$$
$$\omega_e = 2\pi / t_n = 1.06387175271756E\text{+}23 \text{ c/s}$$
$$E_0 = \tfrac{1}{2} J_e \cdot \omega_e^2 = 4.33820131944073E\text{-}17 \text{ J}$$
$$E_1 = \delta KE \cdot (r_e / R_n)^2 = 0 \text{ J} \quad (\delta KE = 0) \quad (E_3 = 0 \text{ J})$$
$$SE_e = 4.33820131944073E\text{-}17 \text{ J}$$

Spin Energy:

$$\begin{aligned} SE &= SE_p + SE_e \\ SE &= |\text{-}4.09355561131267E\text{-}14| + |4.33820131944073E\text{-}17| \\ &= 4.09789381263211E\text{-}14 \text{ J} \end{aligned}$$

Total Energy:

$$E_n = |KE| + |PE| + |SE| = 1.63785606465701E\text{-}13 \text{ J}$$

The Mathematical Laws of Natural Science

5.3.3 Neutron Energy Cell

A neutron energy cell can create neutrons by replicating the conditions in the core of a bright star, only to a much lesser degree. For example, the 100-gram beach pebble shown here possesses 1.37E+12 Joules of neutron [heat] energy; sufficient to fuel:

an average domestic UK household for almost 76 years, or;

an average domestic UK car for almost 100 years.

Because it is 230 million percent efficient, once activated, it will fuel itself. And you can adjust its release rate, or switch it off perfectly safely.

There is sufficient energy in the neutrons of less than one centimetre of the surface of the earth's crust to fuel our energy needs until the next 'Big-Bang'.

The energy a neutron stores at the time of its creation includes the kinetic energy in the electron, the potential energy in the proton-electron pair and the spin energies of the two particles at the time of their union.

Neutrons are the energy source in atom bombs and nuclear reactors, both of which currently rely on the critical mass of the heaviest naturally occurring radioactive element (Uranium). But this release method is both dangerous and expensive. Neutron energy exists in, and can be released safely and cheaply from any matter, including; rock, soil, garbage, plastic, nuclear waste, etc.

The neutron is therefore, not only a key feature in the workings of the universe; it should also be the sole source of energy for any intelligent universal life-form.

5.3.4 Verification

Apparently, when 'Little Boy' was dropped on Hiroshima ≈1.0kg of its mass was destroyed, releasing; **6.3E+13 Joules** of energy (empirical value).

proton separation:
$$C = 4.R_n^2/(r_p+r_e) = 1.65331294837664E\text{-}14 \text{ m}$$

Proton kinetic energy (according to Coulomb):
$$KE_p = k.e^2/C = 1.3954267683677E\text{-}14 \text{ J}$$

neutronic ratio of U_{235}: $\psi = 1.587270761$

1 kg of which contains:
Neutrons; $N_n = 3.66585231725022E+26$
Protons; $N_p = 2.30953181248165E+26$

According to Newton and Coulomb, 'Little-Boy' released:

$$E_T = N_n.E_n + N_p.KE_p = \textbf{6.3264167012986E+13 Joules} \text{ of energy}$$

5.3.5 Internal Stress

When the two particles in a proton-electron pair unite they are both spinning according to the following rules:

The surface velocity of the orbiting electron adjacent to its proton:
$$\omega_e = c/R_n = 1.06387175271756E+23 \text{ °/s}$$
$$v_e \approx \omega_e.(R_n-r_e) = 284361418.5 \text{ m/s}$$

The proton's surface velocity:
$$v_p = r_p .\sqrt{[\, 2.\,|\,KE+PE\,|\, /J_p\,]} = 11062072.34 \text{ m/s}$$

Their rotational velocity ratio is therefore: $v_e:v_p \approx 25.7$

5.3.6 Hypothesis

It is proposed that at the time of their union, this velocity ratio generates an internal stress between the proton and its orbiting electron, which is responsible for a neutron's decay. It is also assumed that excessive neutron-neutron interaction intensifies this stress, decreasing the period over which the particles can exist as a neutron.

6   Proton-Electron Pair

A proton-electron pair is a single proton with a single electron orbiting in a circular path.

A proton-electron pair is cojoined by its quanta opposite electrical charges.

Once trapped by a proton, an electron collects EME from its environment, which it converts to kinetic energy and transmits charge to its proton partner.

Proton-electron pairs collect, generate and radiate all universal EME.

The electron's orbit is circular because it provides its own kinetic energy; i.e. it is not generated by a varying orbital potential energy between it and its proton.

This partnership acts like a generator; a negative electro-magnetic charge orbiting a positive electro-magnetic charge will generate electro-magnetic energy (EME). Only proton-electron pairs emit EME.

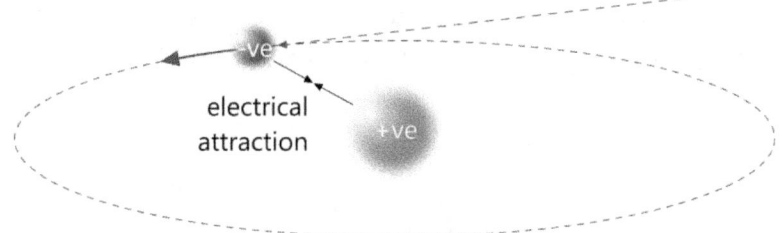

EME is absorbed by orbiting electrons and converted into kinetic energy.

An orbiting electron collects EME from its environment, converts it to kinetic energy, and simultaneously transfers it to its proton partner via their opposite electrical charges.

The proton uses its additional magnetic charge (m_p-m_e) to collect the additional electrical charge its electron partner transfers from its surrounding EME. This additional electrical charge (e') repels protons in the same and adjacent atoms.

Electrical sharing (as opposed to magnetic accruing) is the reason protons can only host one orbiting electron, despite their additional electrical charge (e').

The Mathematical Laws of Natural Science

Proton-electron pairs collect and generate EME, which rises and falls with the kinetic energy in the electron, which rises and falls with the energy in the radiation feeding it. In other words, the EME generated by a proton-electron pair is the same magnitude as the EME feeding it.

According to Newton's laws of orbital motion, an electron's orbital radius decreases with increasing kinetic energy. In other words, atomic structural strength rises with increasing temperature.

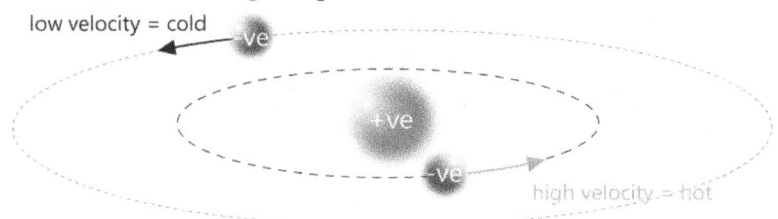

In addition to EME, a proton-electron pair also generates a polar magnetic field - that works contrary to the electrical field - in that it travels from the positive (North) face of the orbital plane and around to the negative (South) face of the orbital plane. This is the magnetism that unites the pair and holds onto adjacent neutrons. It is also the magnetism you see in bar magnets when the atoms in matter are aligned. A proton-electron pair with an attached

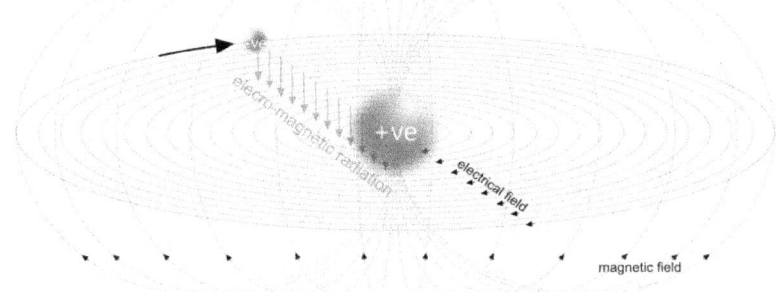

neutron is what we call deuterium.

The relative strengths of the variable electrical charges (e') and the constant magnetic fields in an atom's proton-electron pairs determine whether matter is viscous or gaseous.

Once trapped, an electron will remain in orbit around its proton partner – even inside an atom - until one of the following events occur:

The Mathematical Laws of Natural Science

1) An adjacent atom provides sufficient electrical charge to cause it to swap atoms,

2) Another electron impacts it with sufficient energy to knock it out of its orbit, it is then a free electron,

3) Its velocity falls below which, its proton can no longer hold onto it.

Electrons ejected from a partnership will hold their linear (v) and angular (ω) velocities at the time of ejection in free-flight until affected by magnetism (gravity), interaction with other electrons and/or it is trapped by a proton. What we see in bubble chambers as post-impact spiral paths is simply the result of impacting electrons that can be visualised as spinning billiard balls obeying Newton's laws of motion.

Electrons and protons possess electrical charges of equal and opposite magnitude when at rest ($N_t \approx 1$). The electrical charge in the electron never changes. That of the proton remains constant and equal to that of the electron when it is not part of a proton-electron pair, but varies with electron kinetic energy as soon as it acquires an orbiting electron.

The Mathematical Laws of Natural Science

6.1 Calculations

The orbital properties of a proton-electron pair are related to its temperature

Orbital velocity: $v = \sqrt{[T/X]}$ {m/s}

Orbital radius: $R = X^R/T$ {m}

where: X and X^R are heat transfer coefficients

All the formulas here have been derived from Newton's laws of motion.

According to Coulomb, a proton-electron pair's electrical performance (electrical charge) can be calculated from the formulas listed in the Tables below.

The total energy (including spin) in a proton-electron pair at 300K is:

$E_A = -7.88085938865144E-20$ J

The kinetic energy in an orbiting electron at 300K is:

$KE_A = 1.970214847162860E-20$ J

Sym	Description	units
m_e	electron mass	kg
m_p	proton mass	kg
T	temperature	K

Table 6.1A: *Input Data*

Sym	Formula	Description	units
R	X^R/T	orbital radius	m
e	0	orbital eccentricity	
A	$\pi.R^2$	orbital swept area	m^2
L	$2.\pi.R$	orbital path length	m
K	t^2/R^3	orbital constant of proportionality	s^2/m^3

Table 6.1B: *Electric Orbital Shape*

Sym	Formula	Description	units
v	$\sqrt{[T/X]}$	electron velocity	m/s
t	$2.\pi.R/v$	orbital period	s
a	$-v^2/R$	satellite centrifugal acceleration	m/s^2
F	$-k.(e/R)^2$	proton-electron pair potential force	N
PE	F.R	proton-electron pair potential energy	J
KE	$-PE/2$	satellite kinetic energy	J
F_c	$m_e.a$	satellite centrifugal force	N
E	PE + KE	total energy	J
h	R.v	constant of motion	m^2/s

Table 6.1C: *Electric Orbital Performance*

Page 64

The Mathematical Laws of Natural Science

According to Gilbert & Newton, a proton-electron pair's magnetic performance (magnetic charge) can be calculated from the formulas listed in the following Tables.

Sym	Description	units
m_e	electron mass	kg
m_p	proton mass	kg
T	temperature	K

Table 6.2A: *Input Data*

Sym	Formula	Description	units
R	X^R/T	orbital radius	m
e	0	orbital eccentricity	
A	$\pi.R^2$	orbital swept area	m^2
L	$2.\pi.R$	orbital path length	m
K	$\varphi.(2.\pi)^2 / G.m_p$	orbital constant of proportionality	s^2/m^3

Table 6.2B: Magnetic Orbital Shape

Sym	Formula	Description	units
a	$-G.m_e/R^2$	satellite centrifugal acceleration	m/s^2
v	$\sqrt{|a.R|}$	electron velocity	m/s
t	$2.\pi.R / v$	orbital period	s
F	$-G.m_e.m_p/R^2$	proton-electron pair potential force	N
PE	F.R	proton-electron pair potential energy	J
KE	-PE/2	satellite kinetic energy	J
F_c	$m_e.a$	satellite centrifugal force	N
E	PE + KE	total energy	J
h	R.v	constant of motion	m^2/s

Table 6.2C: Magnetic Orbital Performance

The ratio between the two calculation methods; (magnetic:electrical) is thus;

Difference	Symbol	Ratio	
a:a	φ	4.4074211179233E-40	$.../s^2$
v:v	$\sqrt{\varphi}$	2.0993858906650E-20	.../s
t:t	$\sqrt{\varphi}$	2.0993858906650E-20	.../s
KE:KE	φ	4.4074211179233E-40	$.../s^2$
PE:PE	φ	4.4074211179233E-40	$.../s^2$
F:F	φ	4.4074211179233E-40	$.../s^2$
$F_c:F_c$	φ	4.4074211179233E-40	$.../s^2$
E:E	φ	4.4074211179233E-40	$.../s^2$
h:h	$\sqrt{\varphi}$	2.0993858906650E-20	.../s

Table 6.3: Magnetic:Electrical Performance Ratios

The Mathematical Laws of Natural Science

Note: if we take the force-centre mass of a proton-electron pair according to Newton-Coulomb:
m = $(2\pi)^2$ / G.K = 3.79501207866902E+12 kg
(K = 0.155878745334030 s^2/m^3)
and divide the mass of a proton (the force-centre of a proton-electron pair) by it, we get:
*$\varphi = m_p / m$ = 4.40742111792334E-40; **the coupling ratio!***

Within an atom, the electrical properties dominate the performance of its particles by the coupling ratio (φ). The electrical properties within a proton-electron pair are therefore calculated thus:

Sym	Formula	Description	units
V	PE/e	voltage	J/C
I	e.f	current	C/s
Ω	V/I	resistance	J.s/C^2
ρ	Ω.R	resistivity	J.s.m/C^2
e'	m_e.RC . T/T_n	proton electrical charge	C
μ	m_e.R/e^2	magnetic field	Kg.m/C^2
λ	c/f	EME wavelength	m
A	R	EME amplitude	m
f	v / $2\pi R$	EME frequency	/s
EME	h'/A = ½.m_e.v^2	EME (energy)	J

Table 6.4: Electrico-Magnetic Performance

7 The Atom

Atoms are collections of proton-electron pairs.

The atom according to Newton and Coulomb is a system that works perfectly, it needs no sub-atomic particles to hold it together, and every proton, electron and neutron is identical. Uniqueness and uncertainty are unnecessary. Moreover, apart from Planck's contribution, everything needed to resolve the atom completely and accurately was available before the beginning of the twentieth century.

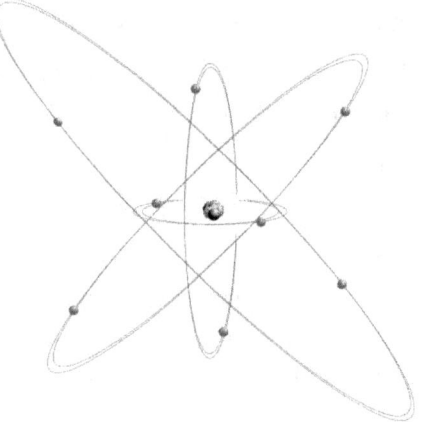

Because all protons repel each other due to their identical electrical polarity, every proton-electron pair within an atom must hold onto at least one neutron isolating it from its neighbouring protons. Proton-electron pairs are the most basic form of matter; hydrogen gas (H), which exist in three forms:

Hydrogen (H); a proton-electron pair.

Deuterium (D); a proton-electron pair with a neutron attached.

Tritium (T); a proton-electron pair with two neutrons attached.

All atoms are created through fusion in the universe's largest massive cold bodies; the ultimate body, the great attractor and galactic force-centres (Chapters 13.1 to .3 respectively). And all neutrons are created by fissionable energy (Chapter 7.7), converting Shell-1 proton-electron pairs into neutrons that are subsequently trapped by adjacent deuterium atoms.

An atom is therefore a collection of deuterium and tritium atoms in which all the proton force-centres are assembled inside a cloud of orbiting electrons, each of which orbits its proton partner.

Whilst natural hydrogen at the surface of a planet comprises mostly lone protons (H+), the majority of hydrogen atoms at the surface of bright stars and planets are proton-electron pairs (H) as they are the residue of total fissionable dissemination.

The Mathematical Laws of Natural Science

It is currently claimed that hydrogen atoms are diatomic; i.e. they exist in pairs (H_2). This is of course impossible for 99.97% of what we refer to as atmospheric hydrogen (H+), as all lone protons possess positive electrical charge, they naturally repel each other. On the other hand, hydrogen atoms (H), which are proton-electron pairs, generate polar magnetic fields (page 62). which means that an hydrogen atom could hold onto a neighbouring hydrogen atom even if neither possessed a neutron. However, you cannot fuse hydrogen atoms, because they possess no neutralising neutrons. You can only fuse deuterium (D) and tritium (T) atoms, which constitute less than 0.03% of all hydrogen atoms. This is the reason we know that it is impossible our sun and its solar system could have accreted from hydrogen gas.

An atom's atomic number (Z) is defined by the number of protons in its nucleus. This number distinguishes its primary character.

Natural atoms have atomic numbers between 1 (hydrogen) and 92 (uranium). The only apparent exception to this rule is technetium ($Z = 43$).

Unnatural atoms, i.e. those with atomic numbers greater than 92 (and 43) can be created artificially but they are very unstable. An unstable atom continuously and regularly ejects its neutronic particles (protons and electrons) or breaks apart due to excess neutron-neutron interaction.

Atoms with an atomic number greater than one are created by fusion; i.e. forcing the nucleus of one atom inside the electron shells of another using magnetic (potential) energy. No two atoms can be fused together unless all the protons in both atoms are protected by at least one neutron. Moreover, fusion can only occur naturally in massive cold bodies. Fusion in stars is impossible for two reasons:

1) stars have insufficient mass to generate the core pressure required, and
2) stars are hot, inter-atomic repulsion prevents fusion from occurring

The total kinetic energy in all the electrons in any atom is its quantity of heat. The highest kinetic energy, is in the electrons orbiting in an atom's innermost shell (shell-1).

All of the proton-electron pairs in any atom collect and radiate EME from their surroundings. The range of visible EME emitted by all of the proton-electron pairs within an atom are its Balmer lines.

7.1 Electron Shells

Every atomic electron orbits its proton partner (not its atomic nucleus).

An electron shell is a specific orbital radius.

There are two orbiting electrons for each specific orbital radius.

The highest energy is generated by the proton-electron pairs in shell-1.

The spacing between orbital shells is equal to the innermost orbital radius.

Orbital electrons are spaced by the electrical balance between the repulsion with other orbiting electrons and attraction with its proton partner.

The orbital performance of each electron is as defined by Isaac Newton and Charles-Augustin de Coulomb.

There are no shell valences.

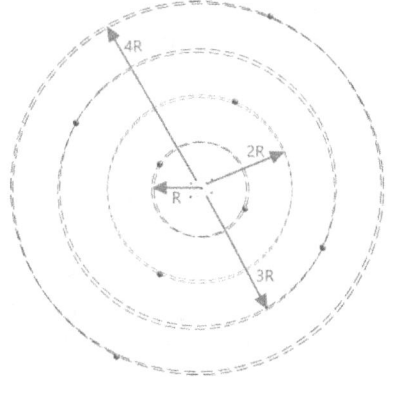

All electron shells are circular, equally spaced, and - apart from the hydrogen atom and the outer shell of atoms with an odd number of electrons - *contain two electrons.* This *statement* is slightly misleading, however; because each electron orbits its proton partner, therefore the two electrons in any given shell cannot follow the same circular path. What this *statement* actually means is that they both orbit their proton partners at the same orbital radius, but at opposite sides of the nucleus.

Shell spacing is defined by a balancing act between the repulsion forces in adjacent electrons and the attraction forces in their proton partners, in exactly the same way as Newton described the balance between centrifugal force and potential force in an orbiting satellite and its force-centre.

When an inner-shell electron is lost, all other electrons redistribute to fill the gap. Redistribution occurs automatically as electrical forces balance. Shell spacing is equal to the radius of the innermost shell (shell-1).

The Mathematical Laws of Natural Science

The orbital radii of all atomic electrons decrease simultaneously as the surrounding EME rises, and increase simultaneously as the surrounding EME falls.

Shell radius is determined by electron velocity, which is determined by its kinetic energy, which is determined by the EME it collects from its surroundings. The greater an electron's velocity, the smaller its orbital radius, and vice-versa. An electron cannot collect more EME than its balancing forces permit. Shell-1 holds the most energy and shell-N (outermost shell) holds the least.

We measure electron kinetic energy as temperature.

The first law of thermodynamics limits the highest temperature of any atom - that of its shell-1 electrons - to its environmental EME, which means that the temperature of all its other electron shells must decrease with increasing radius. This is the reason the specific heat capacity of elements decreases with increasing atomic number. The upshot of which, means that at any given environmental temperature, metals with a higher atomic number (Z) feel colder to the touch.

7.2 Nucleus

The lattice structure of an atomic nucleus is replicated in the atoms of same-element matter in both viscous and gaseous states.

The nucleus of an atom contains proton partners together with their attached neutron(s) that are held by the magnetic field energy generated by the proton-electron pairs.

Their positive electrical charges (e') prevent nucleic protons - and their neutron partners - from sitting together in the atomic nucleus. Instead, they will organise themselves, spaced apart, by balancing their electrical charges - both repulsion and attraction - within the innermost orbiting electrons, settling at their lowest energy condition, creating each atom's unique lattice structure.

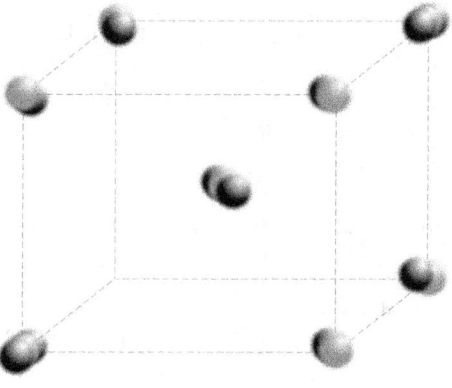

Nucleic organisation (lattice arrangement) is defined by the electrical isolation of its proton's positive charges, which is determined by its attached neutrons. Neutrons are therefore responsible for an atom's lattice structure. The arrangement of any atomic nucleus is replicated in the lattice structure of the atoms of elemental matter in both viscous and gaseous conditions.

Because of the replication of atomic nucleus and elemental lattice structures, their mathematical relationships are identical.

Regardless of the nucleic pattern, structural integrity generally reduces with increasing nucleic size (i.e. as the atomic number increases), making larger atoms generally (but not necessarily) more unstable due to the greater potential for neutron-neutron interaction, resulting in radioactivity. Technetium appears to be a special case in which the nucleic arrangement of 43 protons is especially vulnerable to neutron decay.

7.2.1 Nucleic Structure

The lattice structure (ζ) of all matter of any atom is unique to its neutronic ratio and is replicated in elemental matter in both viscous and gaseous states. This uniqueness implies that the current arrangements; close-packed hexagonal, face-centre cubic, tetrahedral, etc. are over simplified. Every lattice structure appears to be unique.

7.2.2 Nucleic Property Table

The following Table lists the nominal neutronic ratio (ψ) and lattice factor (ζ) of the 92 natural elements.

The <u>underlined</u> elements are naturally radioactive.

The **bold** elements are the noble gases (see below).

The **noble gases** appear to be dominated by the number **9**;
    2, 9, 18, 36, 54 & 86

i.e. when the factor '$\Gamma = 9.(\psi-1)$' is at or closest to an integer; (0) to (5). This integer value appears to replace neon as a noble gas with fluorine. The reason for this anomaly may well be that the Periodic Table may not be as definitive as currently believed.

The Mathematical Laws of Natural Science

Element	Z	ψ	Γ	Noble	ζ
Hydrogen	1	0	0.0715		3.866
Helium	2	1	0.0117	0	9.308
Lithium	3	1.31	2.823		1.6824
Beryllium	4	1.25	2.2774		1.8068
Boron	5	1.16	1.46		1.7816
Carbon	6	1	0.0161		1.8076
Nitrogen	7	1	0.0086		5.5909
Oxygen	8	1	0.0001		5.902
Fluorine	9	1.11	0.9984	1	6.3248
Neon	10	1.02	0.1617		8.8979
Sodium	11	1.09	0.8098		2.4543
Magnesium	12	1.03	0.2288		2.6621
Aluminium	13	1.08	0.6795		2.3381
Silicon	14	1.01	0.055		2.3207
Phosphorus	15	1.06	0.5843		3.7295
Sulphur	16	1	0.0366		3.5339
Chlorine	17	1.09	0.7692		4.7768
Argon	18	1.22	1.974	2	6.7282
Potassium	19	1.06	0.5202		2.6535
Calcium	20	1	0.0351		2.5189
Scandium	21	1.14	1.2668		2.4423
Titanium	22	1.18	1.582		2.576
Vanadium	23	1.21	1.9336		2.744
Chromium	24	1.17	1.4985		3.068
Manganese	25	1.2	0.7777		3.3648
Iron	26	1.15	1.3317		3.0942
Cobalt	27	1.18	1.6444		3.1758
Nickel	28	1.1	0.8657		3.18
Copper	29	1.19	1.7212		3.3402
Zinc	30	1.18	1.6155		4.2666
Gallium	31	1.25	2.2422		3.218
Germanium	32	1.27	2.4216		2.929
Arsenic	33	1.27	2.4332		4.5338
Selenium	34	1.32	2.9012		4.2745
Bromine	35	1.28	2.5467		5.5397
Krypton	**36**	**1.33**	**2.9495**	**3**	**7.4403**

Table 5.1: Noble Gases & Lattice Factors

The Mathematical Laws of Natural Science

Element	Z	ψ	Γ	Noble	ζ
Rubidium	37	1.31	2.7895		3.3437
Strontium	38	1.31	2.7521		3.157
Yttrium	39	1.28	2.5167		2.7391
Zirconium	40	1.28	2.5254		2.741
Niobium	41	1.27	2.3941		2.8527
Molybdenum	42	1.28	2.5586		2.9933
Technetium	43	1.3	2.7013		3.1672
Ruthenium	44	1.3	2.6734		3.2417
Rhodium	45	1.29	2.5811		3.3868
Palladium	46	1.31	2.8213		3.6129
Silver	47	1.3	2.6556		3.8601
Cadmium	48	1.34	3.0771		4.9376
Indium	49	1.34	3.089		3.6317
Tin	50	1.37	3.3642		3.4052
Antimony	51	1.39	3.4853		3.871
Tellurium	52	1.45	4.0846		4.3626
Iodine	53	1.39	3.5498		5.8039
Xenon	54	1.44	3.984	4	7.3059
Caesium	55	1.42	3.748		3.6951
Barium	56	1.45	4.0704		3.2524
Lanthanum	57	1.44	3.9324		3.0566
Cerium	58	1.42	3.7421		3.1312
Praseodymium	59	1.39	3.4944		3.1121
Neodymium	60	1.4	3.636		3.2778
Promethium	61	1.38	3.3934		3.3307
Samarium	62	1.43	3.8265		3.928
Europium	63	1.41	3.7091		3.7538
Gadolinium	64	1.46	4.1133		3.3344
Terbium	65	1.45	4.005		3.382
Dysprosium	66	1.46	4.1591		3.6671
Holmium	67	1.46	4.1548		3.6406
Erbium	68	1.46	4.1372		3.6039

Table 5.1 (cont.): Noble Gases & Lattice Factors

The Mathematical Laws of Natural Science

Element	Z	ψ	Γ	Noble	ζ
Thulium	69	1.45	4.0349		4.0733
Ytterbium	70	1.47	4.248		4.3955
Lutetium	71	1.46	4.1789		3.5005
Hafnium	72	1.48	4.3113		3.4374
Tantalum	73	1.48	4.3086		3.394
Tungsten	74	1.48	4.3601		3.5064
Rhenium	75	1.48	4.3448		3.5612
Osmium	76	1.5	4.5272		3.7649
Iridium	77	1.5	4.4669		3.9183
Platinum	78	1.5	4.509		4.0626
Gold	79	1.49	4.4392		4.3466
Mercury	80	1.51	4.5664		6.8709
Thallium	81	1.52	4.7093		4.7566
Lead	82	1.53	4.7415		4.4932
Bismuth	83	1.52	4.6605		4.4941
Polonium	84	1.49	4.3909		5.0587
Astatine	85	1.47	4.2339		6.0774
Radon	**86**	**1.56**	**5.0244**	**5**	**7.789**
Francium	87	1.56	5.071		4.9528
Radium	88	1.57	5.1162		4.217
Actinium	89	1.55	4.9579		3.6903
Thorium	90	1.58	5.2038		3.3745
Protactinium	91	1.57	5.1429		3.783
Uranium	92	1.59	5.2854		3.9499

Table 5.1 (cont.): Noble Gases & Lattice Factors

7.3 Neutronic Ratio

All elements comprise collections of deuterium (D) and tritium (T) atoms. The neutronic ratio (ψ) of an atom is the ratio of these atoms;

$\quad$ T:D = RAM/Z - 1

Therefore, the 'neutronic ratio can theoretically be between: $1 < \psi < 2$

However, whilst the theoretical upper limit for ψ is 2, neutron-neutron interaction in atoms with a neutronic ratio (ψ) greater than $1\frac{2}{3}$ will spontaneously self-destruct.

A neutronic ratio of 1.6 is the actual practical upper limit.

Atoms with a neutronic ratio (ψ) greater than 1.5 will continually eject neutrons as alpha and beta-particles, making the atoms radioactive.

Over time, atoms naturally try to achieve $\psi = 1$, which is their most stable form. They eventually achieve this by ejecting the surplus neutrons in their tritium atoms as alpha and beta-particles. The rate at which this occurs is referred to as the 'half-life' of the atom.

To summarise, for any atom:

$\quad \psi > 1\frac{2}{3}$: (impossible)

$\quad \psi > 1.6$: self-destruct (most unstable)

$\quad 1.5 < \psi \leq 1.6$: radioactive (unstable)

$\quad 1.0 < \psi \leq 1.5$: chemically reactive

$\quad \psi = 1.0$: contains 100% deuterium atoms (chemically inactive)

Because chemical reactions will not occur between atoms with a neutronic ratio of 1, neutrons are not only responsible for all universal energy, they are also responsible for all chemical reactions; organic and inorganic.

Our universe only exists because of its neutrons.

7.4 Ion

Ions are atoms with the same atomic number (Z) but possess an electrical charge owing to unequal proton-electron pairing.

Positive ions (atoms that have lost electrons) possess a positive electrical charge. Negative ions (atoms with additional electrons) possess a negative electrical charge. Negative ions are far less common than positive ions.

Only a few atoms exist naturally as negative ions and they are all non-metals, except for two, which are semi-metals:

> One additional electron (Group VIIA):
>
> Fluorine (9_N), Chlorine (17_N), Bromine (35_N), Iodine (53_N)
>
> Two additional electrons (Group VIA):
>
> Oxygen (8_N), Sulphur (16_N), Selenium (34_N), Tellurium (52_N)
>
> Four additional electrons (Group IVA):
>
> Carbon (6_N), Silicon (14_N).

Any atom can become a positive ion simply by losing one or more of its electrons from impact with free electrons or a strong external positive electrical charge.

Negatively charged ions are a little more difficult to understand. Additional electrons need to be trapped by the positive charge in protons that do not exist in the nucleus: this shouldn't be possible. However, the nucleic structures of the above non-metal atoms probably have at least one exposed proton that is not protected by a neutron and this means that the additional electro-magnetic electrical charge (e') generated in it is available to trap passing free electrons

The Mathematical Laws of Natural Science

7.5 Isotope

Isotopes are atoms with the same atomic number (Z) but with varying atomic mass because of unequal proton-neutron pairing. For example; an atom of iron, with 26 protons (Z=26) and 26 neutrons (N=26) is an isotope of 52. However, in nature, most iron atoms have more than 26 neutrons, each of which is given its own isotope, e.g. 57, 59, etc.

An atom's isotope value is the same as its RAM: RAM = N + Z

In all same element matter, its atomic isotopes will vary, some more unstable than others. Whilst all of them will have different half-lives, they would all eventually achieve an isotope of 2.Z (ψ=1) given sufficient time. Luckily, however, fission in bright stars and planets, and fusion in cold dark bodies together with the limiting universal period of less than 32 billion years means that chemical reactions will continue with each new universal period.

In other words, because the half-life of most elements is such that they will retain excess neutrons longer than a universal period, chemical reactions will continue into a subsequent universal period.

The Mathematical Laws of Natural Science

7.6 Fusion

Atoms are created by forcing the nucleus of one atom inside the electron shells of another.

Fusion only occurs naturally in cold bodies with sufficient mass to generate the necessary core pressure.

Gaseous atoms cannot be fused due to electrical repulsion between protons.

Fusion is the union of proton-electron pairs and/or atoms to create a different element. It is accomplished by applying sufficient force to push the nucleus of one atom inside the electron shells of another.

Atomic fusion can only occur naturally inside cold bodies with sufficient mass to generate the necessary core pressure. This is only possible inside galactic force-centres, the great attractor and the ultimate body, because they are sufficiently massive and they are cold.

When atoms or proton-electron pairs are fused together, a small amount of EME will be released as their electrons rearrange themselves into shells. The energy released, however, will be considerably less than that required to fuse the atoms.

For example, @ 30°C:

> The kinetic energy in a carbon atom = $7.29997E-20$ J
> The kinetic energy in an iron atom = $1.26627E-19$ J
> Individually, they generate a total of: $1.99626E-19$ J
> United (as Germanium), they would generate: $1.34614E-19$ J

Releasing: $6.50124E-20$ J

> The potential energy in a carbon atom = $-1.45999E-19$ J
> The potential energy in an iron atom = $-2.53253E-19$ J
> The energy required to unite these two would be: $-3.8598E-19$ J

Representing a nett energy input of: $-3.20968E-19$ J

Therefore, it is necessary to input $3.2E-19$ Joules of potential energy in order to release $6.5E-20$ Joules of kinetic energy; in other words, in this case you need to input five times as much energy as you're releasing.

The Mathematical Laws of Natural Science

Fusion requires the input of energy; it does not generate energy. That is why Hades is cold and more than 50 years of trials have yet to produce a fusion reactor.

It is also important to understand that the release of energy due to fusion is a single instantaneous event, once accomplished, no more energy will be released.

Refer to Chapter 11.4.4 for an assessment of the potential for fusion in the cores of our earth and our sun.

The Mathematical Laws of Natural Science

7.7 Fission

Fission is the splitting of neutrons into their component parts, releasing their stored energy; 1.63785606465701E-13 Joules for each neutron split.

The ejection of neutrons is called radioactive decay, and the time over which it occurs naturally is referred to as its half-life. Neutron energy is released either as:

> *electro-magnetic energy* if the neutron decays into a proton-electron pair but is unable to escape from the atomic nucleus. In this case, the energy is released as EME:
>
> EME = $E_n/2$ = 4.09355561131267E-14 J
>
> and the atom concerned will have become a different element (Z+1 for each trapped proton-electron pair).

Or;

> *kinetic energy* if the neutron's particles are released from the atom. In this case, the proton and the electron will be ejected at velocity:
>
> v = $\sqrt{[2.E_n / (m_p+m_e)]}$ = 9891525.10667846 m/s (3.3% of light-speed)
>
> the proton of which will impact neighbouring neutrons splitting them into their component parts (a proton and an electron), the protons of which will then split other neutrons; a chain reaction. An ejected proton will not impact a neighbouring proton because of their similar electrical polarity.

Fission can only be released naturally by raising the temperature of a sufficiently massive celestial body's core atoms to neutronic condition through planetary spin. This is the energy released in all bright stars and planets.

The release of neutrons can be initiated artificially by raising the temperature of an atom's two proton-electron pairs in shell-1 to the neutronic temperature (T_n); causing both of them to become neutrons. If this is continued, the atom's neutronic ratio will quickly exceed 1.6, at which time, the atom will eject neutronic particles (protons and electrons), the kinetic protons of which will impact and split other neutrons releasing their energies; a chain reaction. Just one atom in any mass will result in the release of much more neutron energy than that required to achieve $\psi>1.6$ in a single atom. It is this process which occurs in the core of bright stars and planets that have been heated through planetary spin.

The heat we feel from bright stars and planets is fissionable energy released as EME, and the hydrogen atmosphere created at their surface is the residue of total neutronic dissemination; the breaking down of core atoms into proton-electron pairs.

Whilst neutrons are continually ejected by all atoms with a neutronic ratio greater than '1', we refer to those with a neutronic ratio greater than '1.5' as unstable because they represent a danger to life-kind.

Neutron-neutron interaction of a mass of atoms with a neutronic ratio close to or greater than 1.6 will increase the risk of a chain reaction causing it to break apart. The critical mass of such matter is that at which it will become impossible to prevent self-destruction. If its critical mass is achieved quickly enough, the ejected atoms will have nowhere to go so the matter will break apart instantly. This is what occurs in an atom bomb.

Whilst our knowledge today allows us only to exploit such matter using the critical mass of radioactive matter, neutron energy could be released from any matter in a controlled manner if processed correctly. I.e. neutron energy could be safely released from; metals, plastics, earth, rocks, waste, etc., and also from nuclear waste.

Radioactivity can, and should be regarded as a friend, not an enemy.

7.8 Half-Life

Every proton-electron pair inside an element with an atomic number greater than one must have a neutron partner in order for the atom to exist as such.

The term "half-life" of an atom does not mean that an atom will exist twice as long as its half-life.

All elements want to revert to their most stable structure; $\psi = 1$, therefore, over time all excess neutrons ($\psi>1$) will eventually be ejected from all elements. The time over which an atom loses half of its excess neutrons is described as its 'half-life'. This period can range from seconds to billions of years.

For example, if one in every thousand carbon atoms possess two tritium atoms, the time taken to reduce this ratio by half is termed its half-life. It would appear (currently claimed) that the period over which this reduction occurs is fixed and constant, irrespective of conditions.

Whilst the mechanism behind the creation of neutrons is now known, their demise, which is apparently time-driven, has yet to be defined mathematically. However, it is expected that a neutron's decay-rate is linked to both; the internal stress generated by the difference between the curvilinear surface velocities (x25.7; chapter 5.3.5) of the electron and its proton due to spin at the time of their union; and the magnetic energy generated due to neutron-neutron interaction within a nucleus.

7.9 Calculations

An atom is a collection of proton-electron pairs (with neutrons attached).
An atomic shell is an orbital radius.
There are two orbits for each orbital radius.

This is why we say each atomic shell holds 2 electrons. Both electrons in any shell orbit their proton partner, so they will have different centres; they are offset. Both electrons in any shell will have the same performance (velocity, energy, force, etc.).

The number of orbital shells in any atom: $N° = INT([Z-1]/1.9) + 1$

If $\underline{T}$ is the measured temperature of the atom;
    the innermost orbital shell radius: $R_1 = X^R/\underline{T}$ {m}
    and the orbital radius of any shell: $R_N = N°.R_1$ {m}
    and its temperature: $\underline{T}_N = X^R/R_N$ {K}

You may then use the formulas provided in chapter 6.1 to define the performance of all the proton-electron pairs in the atom. Because electrical charges are dominant, you must define the atom's performance using those in Tables 6.1

7.9.1 Atomic Density

The density of an atom varies with temperature ($\underline{T}$). The higher the temperature, the greater the atomic density. It may be calculated as follows:

$$\rho = {}^4/_3 \pi . R_N^3 \qquad \{kg/m^3\}$$

For example; the density of iron is 7870 kg/m³

whilst the density of its atom:
    @ 273.15 K: $\rho = 0.083888668$ kg/m³
    @ 12,412 K: $\rho = 7870$ kg/m³

This 'electron-clouding' allows lone protons (and positive ions) to share spare electron charge capacity in neighbouring atoms at low temperatures, resulting in viscosity and chemical bonding.

Whilst chemical and inter-atomic bonding weaken with increasing temperature as orbital radii reduce, atomic strength and density will rise.

The Mathematical Laws of Natural Science

7.9.2 Specific Heat Capacity

The specific heat capacity of an atom is the sum of the kinetic energy in all its proton-electron pairs relative to its mass and its measured temperature ($\underline{T}$).

The plot below shows the calculated values for specific heat (ΣKE) for all atoms from Z=4 to 92 compared to the documented values that have been taken from various sources and which are subject to experimental error.

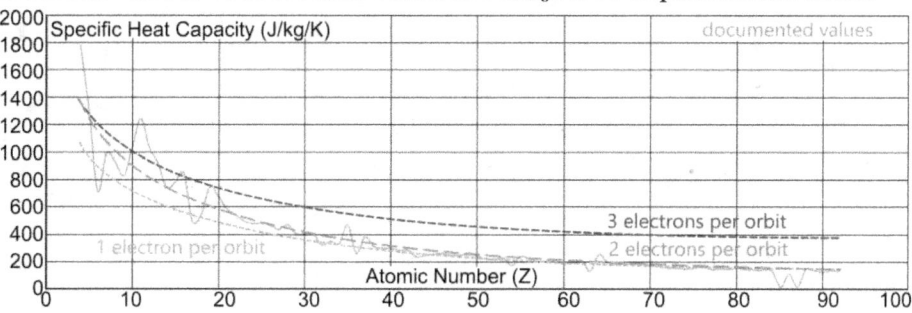

As can be seen, the relationship between the documented values and those calculated for 2 electrons per shell are similar, whereas the calculated values for 1 and 3 electrons per shell are obviously incorrect. This is how we *know* that each shell contains two electrons.

The specific heat capacity of any atom may be calculated thus:

$$\text{SHC} = \Sigma KE / Y.m.\underline{T} \qquad \{J / kg.K\}$$

where:
ΣKE = the sum of kinetic energies in all the atom's proton-electron pairs
$\underline{T}$ = the measured temperature of the atom
Y = temperature constant
m = atomic mass (including electrons and neutrons)

The Mathematical Laws of Natural Science

8 The States of Matter

All matter can exist only in either viscous or gaseous states.

The melting point of matter is the temperature beyond which it can no longer hold its shape under its environmental conditions; ambient pressure and gravitational acceleration.

Matter is the term used to describe a collection of atoms, and same-element matter describes a collection of atoms with the same atomic number. All matter can only exist in either viscous or gaseous state. Gas-transition is the term used for the temperature at which matter changes between these states.

The only difference between solid and liquid matter is its ability to retain its shape under the forces of gravity and external pressure.

Melting point is therefore a misnomer, as it will occur at different temperatures dependent upon environmental conditions.

Fluid is also a misnomer, as it unites two disparate states; viscous and gaseous.

Boiling point is not necessarily the same as gas transition temperature. Most elemental matter evaporates before achieving gas-transition. A vapour is a liquid in suspension in a gas such as an atmosphere. The properties of a vapour are dependent upon the element's surface tension and its density.

8.1 Lattice Structure

The atomic arrangement in same element matter is replicated from its nucleic structure, in both viscous and gaseous states. The mean atomic spacing between adjacent atoms may be calculated thus: $d = \sqrt[3]{[m_a/\rho]}$

But the actual spacing between any two adjacent atoms will vary with direction, some greater than 'd' and others less. The mean magnetic force, which depends upon inter-atomic spacing, may be defined by a lattice factor (ζ) thus: ζ/d

Refer to chapter 7.2.2 for the numerical values of 'ζ'.

The Mathematical Laws of Natural Science

8.2 Inter-Atomic Forces

Adjacent atoms are attracted and repulsed by opposing magnetic field and electrical charge forces, that are together responsible for matter's viscous or gaseous state.

Every proton-electron pair in every atom generates a magnetic field force that holds adjacent atoms together, and a variable positive electrical charge (e'), which forces them apart. The highest of these forces are generated by the proton-electron pairs in shell-1. PE_1 is therefore the dominant potential energy.

The 'PVRT' formula applies equally to viscous and gaseous matter:

$$p = R_i \cdot \underline{T_1} \cdot \rho / (RAM/1000) \qquad \{N/m^2\}$$

But we can also calculate it like this ...

$$p = -PE_1 / Y.d^3 = k_B \cdot \underline{T_1} / d^3 = \rho.k_B.\underline{T_1} / m_a \qquad \{N/m^2\}$$

$$p = m_e.c^2 / Y.d^3 \cdot \underline{T_1}/\underline{T_n} = PE_n / Y.d^3 \cdot \underline{T_1}/\underline{T_n} \qquad \{N/m^2\}$$

... because they all generate exactly the same result (alter RAM to RAM.N and m_a to m_M for molecular calculations).

As force is pressure multiplied by area, the electrical force of repulsion between adjacent atoms may be calculated thus ...

$$F_e = p.d^2 = PE_1 / Y.d = k_B.\underline{T_1} / d \qquad \{N\}$$

all of which give exactly the same result,

... and inter-atomic force of attraction may be calculated thus:

$$F_m = \mu_o.\xi_m/\zeta^3 \cdot I_1^2 \cdot (2\pi)^2 \cdot (R_1/d)^3 \cdot RAM.m_n/m_p = \rho.h_e^2/\zeta^3 \qquad \{N\}$$

where (subscript-1 refers to shell-1):
ρ = matter density (gaseous or viscous)
d = mean inter-atomic spacing ($\sqrt[3]{[m_a/\rho]}$)
m_a; m_p; m_n = mass of; atom, proton, neutron respectively
$\underline{T_1}$ = measured temperature (T_j) of the matter
PE_1 = potential energy & PE_n = neutronic potential energy
ζ = lattice factor

Both of these forces (F_m & F_e) are equal at gas transition temperature:

viscous matter: $F_m > F_e$
gas-transition: $F_m = F_e$
gaseous matter: $F_m < F_e$

The Mathematical Laws of Natural Science

The electrical repulsion force between adjacent atoms is equal to the electrical attraction force between shell-1 electrons and their proton partners when e' = e, which occurs at the following temperature:

$$F_e = k.(e/d)^2 = k.(e'/d)^2 \qquad @ \; \underline{T} = 339468.852842837 \; K$$

This plot shows the transition from viscosity to gaseous for O_2.

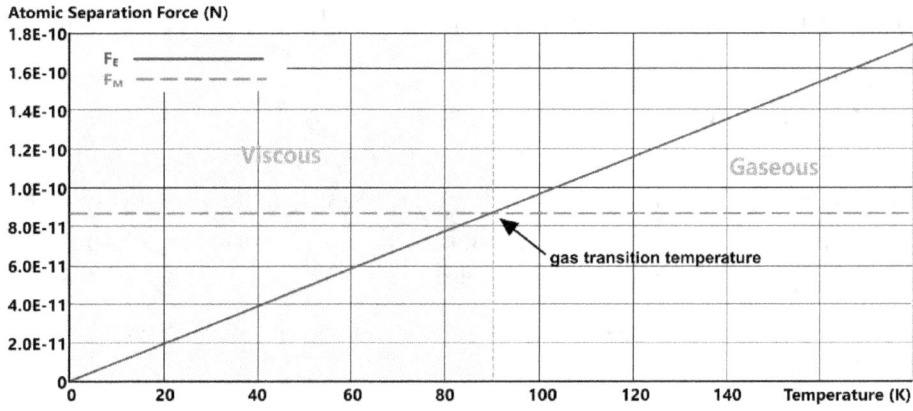

The gas transition temperature of any element can be calculated like this:

$$\underline{T_1} = d.F_m/k_B \, / \, \zeta^3 = (\rho.h_e^2 \, / \, d^2.\zeta^3 + p_{atm}).(RAM/1000) \, / \, R_i.\rho \qquad \{K\}$$

where F_m includes atmospheric force

which occurs when atomic electrical and magnetic forces are equal; $F_m = F_e$
The temperature of any element can be calculated like this:

$$\underline{T_1} = d.F_e/k_B \qquad \{K\}$$

The following image shows a comparison between various documented gas-transition temperatures and the calculated value.

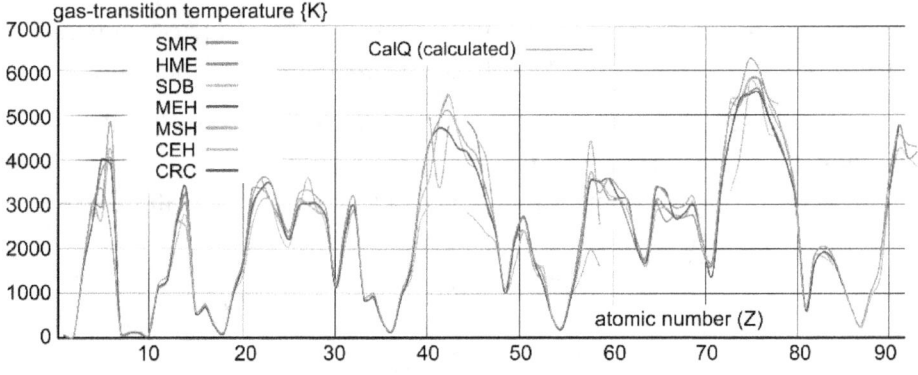

Refer to Appendix A-2 for the publication sources of the documented values.

8.3 Viscous Matter

Viscosity is the consequence of plastic slip between adjacent inter-atomic planes.

Solidity refers to high-viscosity matter, in which plastic slip is difficult to induce.

Liquidity refers to low viscosity matter, in which plastic slip occurs naturally due to gravitational force.

If a proton-electron pair's electron is orbiting slowly (low temperature), the proton's electrical charge (e') will be low, and the electrical repulsion force (F_e) between adjacent atoms will be less than the magnetic attraction force (F_m). In this case, matter is viscous (solid or liquid); the lower the atom's temperature, the more viscous the matter.

Because magnetism is constant at any temperature (chapter 4.2.1), the magnetic field force also remains constant at elevated temperatures.

8.3.1 Electron Clouding

Electron clouding (sharing) is the reason viscous matter holds together; i.e. most electron shell radii are considerably greater than nucleic separation.

For example; the radius of the outer shell of the iron atom at 300K is 7.611E-08 m, whilst inter-atomic separation at the same temperature is 2.2815E-10 m (chapter 7.9.1).

As temperature rises, shell radii shrink, reducing electron clouding and increasing the electrical charge repulsion in the nucleic protons, which together weakens inter-atomic bonding. The increase in electrical charge repulsion is what causes matter to soften with increasing temperature.

Atoms with identical atomic numbers will naturally attract atomically due to their identical atomic structures. In other words, if two molten substances are flowed together, only those with the same nucleic structures will bond atomically.

The Mathematical Laws of Natural Science

8.3.2 Surface Tension

As explained in chapter 8.4.2, all matter is held together by internal pressure; 'p', therefore, all of its properties are based upon this pressure.

Surface tension is the linear force holding adjacent atomic planes together. It can be calculated thus;

$\gamma = p.Y.d$ {N/m}

By way of comparison the surface tension of liquid mercury @ 293K:
is said to be; $\gamma = 0.465$ N/m

whereas the calculated value:
is; $\gamma = 0.4527$ N/m

8.3.3 Elastic Moduli

There are a number of elastic moduli, Tensile, Bulk, Shear, etc. The reason they are elastic is because they assume no plane slippage; i.e., deformation under load will recover completely. The calculated value is purely theoretical, as it assumes a force applied normal to a perfectly aligned crystal (h), which means it will be higher than the tested value.

$h = d.(1-\sin(60°))$ {m}

Tensile modulus can be calculated as follows:

$E = p \cdot (d/(h.\sin(60°)))^3$ {N/m²}

By way of comparison E of iron @ 293K:
is said to be; $E = 1.965E+11$ N/m²

whereas the calculated value:
is; $E = 2.182E+11$ N/m²

tensile modulus

The discrepancy is due entirely to matter condition. Even perfectly aligned crystals prior to test commencement, will not remain as such under tension. It will disrupt, most of which will rotate, resulting in shear, which is the reason that the theoretical value will be greater than the empirical value.

Shear modulus: $G = E / 2.(1+\nu)$ {N/m²}
Bulk modulus: $K = E / 3.(1-2.\nu)$ {N/m²}

The Mathematical Laws of Natural Science

8.3.4 Viscosity

Viscosity is the term we use to describe the slippage between matter's adjacent atomic planes, which are held together by the internal pressure; 'p' (see chapter 8.4.2; p_2)

Dynamic viscosity is calculated thus: $\mu = p / \Gamma.f$ {N.s/m² = kg/m/s}

Kinematic viscosity is calculated thus: $\nu = \mu/\rho$ {m²/s}

By way of comparison the dynamic viscosity of liquid mercury @ 293K:
   is said to be; $\mu = 0.001526$ kg/m/s

whereas the calculated value:
   is; $\mu = 0.00165282$ kg/m/s

The only difference between liquid and solid matter is the magnitude of the force needed to slip its atomic planes. As its temperature falls, the internal pressure (p) reduces, raising the force required to induce slippage, thereby increasing its viscosity.

8.3.5 Stress

In viscous matter, atomic clouding holds adjacent atoms together.

Internal stress in metals is the direct pressure between adjacent atomic planes (see chapter 8.4.2).

   $p = PE_1 / Y.d^3$ {N/m²}

Yield stress is that which opposes this internal pressure, and can be calculated as follows:

   $\sigma_y = p$ {N/m²}

By way of comparison the yield stress of iron @ 293K:
   is said to be; $1.5E+08 < \sigma_y < 5.52E+08$ N/m²

whereas the calculated value:
   is; $\sigma_y = 3.40561E+08$ N/m²

The calculated value is purely theoretical, because it assumes perfect crystal alignment. It is expected, therefore, to be towards the high end of the empirical [documented] range.

8.3.6 Electricity

Every atomic particle possesses an electrical charge. Because electrical charge is polar, it attracts and repels according to the polarity of other electrical charges. This is why electricity is shared between particles; their opposite or similar polarities balance. We see this in electrical circuits; at a given voltage; a single light-bulb will be brighter than multiple light-bulbs in the same circuit.

Electricity is the movement in electron charge(s). Within atoms, this movement is contained in the electron's orbits. If an external electrical charge is sufficient to pull electrons from their orbits, electrons will flow between atoms; through matter.

The higher the temperature of the matter (its proton-electron pairs), the faster the orbiting electrons, the greater the potential energy, and therefore the harder it will be to pull electrons from their orbits; electrical resistance increases with rising temperature.

Electrical energy flows from negative to positive.

8.3.7 Magnetism

It is impossible that an atomic particle with electrical charge cannot also possess magnetic charge. Electricity and magnetism are inextricably linked. Therefore, the magnetic (and electric) charges must be responsible for generating magnetic fields.

Every atomic particle possesses a constant magnetic charge. Because magnetic charge is non-polar, it only attracts all other magnetic charges; it does not repel. This is why magnetism accrues. We see this in celestial bodies; potential energy (gravitational attraction) is greater between bodies containing a larger number of atomic particles than those with less.

Attraction occurs between all non-polar magnetic charges throughout the universe. But the potential energy radiated by a magnetic charge will be distributed over the surface area at a given distance. This is why non-polar magnetic charge attraction (gravity) *appears* to diminish with the square of the distance between bodies.

Because proton-electron pairs generate magnetic fields via their electrical and magnetic charges, magnetic fields are polar, which is why EME's electrical and magnetic energies are polar.

Magnetic EME field energy is not all-pervasive, it has a limited field of influence, but is much stronger than the radial field emanated by a magnetic charge. It flows from the positive face of the proton-electron pair's orbital plane towards the negative face. This is what we see in the pattern generated by iron filings around bar-magnets.

The Mathematical Laws of Natural Science

8.4 Gaseous Matter

If the electron in a proton-electron pair is orbiting quickly (high temperature), the proton's electrical charge (e') will be high, and the electrical repulsion force (F_e) between adjacent atoms will exceed the magnetic attraction force (F_m). In this case, matter will exist as a gas; the greater its atom's temperature, the higher will be its pressure (when contained).

Because proton electrical charge varies [linearly] with temperature, inter-atomic repulsion forces also vary with temperature.

Gaseous atoms are those that repel each other due to the electrical charge (e') in their nucleic protons. Whilst magnetic attraction occurs between adjacent atoms at temperatures a little above gas transition, this attraction reduces rapidly with the cube of their separation distance (refer to Chapter 8.2).

The repulsion forces between adjacent atoms generate the pressure in enclosed gases. Their magnetic attraction is responsible for the drag induced in bodies passing through gaseous matter. Raise the temperature of atmospheric gases and drag will decrease.

8.4.1 Partial Pressures

At first sight, Dalton's law and partial pressure theory appear to conflict.

Dalton's law states that each gas in a mixture of different gases will be evenly distributed within its container independently of all the other gases in the mixture.

Partial pressure theory states that the total pressure of a gas mixture is the sum of the pressures of each individual gas.

If as Dalton states; each gas in a mixture of gases should be treated independently, the pressure in a container of mixed gases should be the maximum individual pressure. But this is not the case; partial pressure theory tells us that the total pressure is the sum of the individual gases. This anomaly may be explained as follows:

The positive electrical charge in all atomic protons (e') will repel all other atoms, irrespective of atomic number, but such repulsion is random.

All same-element atoms will repel each other equally according to their respective lattice structures, and because all energies (and forces) naturally

The Mathematical Laws of Natural Science

settle at their minimum energy level, they will distribute evenly and equally throughout their container, maximising their relative distances (Dalton).

This is why; the total pressure in a container of mixed gases is the sum of all the individual pressures yet same-element gases must be treated individually.

It is interesting to note, that this exclusive interaction between disparate atoms in a gaseous state also applies when viscous, and the reason same element atoms collect together as a liquid.

8.4.2 PVRT

The PVRT formula applies to all matter in both gaseous and viscous states.

Confirmation of this atomic model can be deduced from the ease with which the well-known and accepted gas-pressure calculation formula:

$$p = R_i/(N.RAM/1000) \cdot \underline{T}.\rho = R_a \cdot \underline{T}.\rho \qquad \{N/m^2\}$$

can be replaced with a formula using Newton's and Coulomb's formulas thus;

$$p = PE_1/Y \cdot \rho/m_M \qquad \{N/m^2\}$$

where; PE_1 is the potential energy in the atom's proton-electron pair(s) in shell-1, ρ is the gas density and m_M is the molecular mass

The following Table shows a calculation comparison for our atmospheric gases @ 300K, using three different calculation methods, confirming that the Newton-Coulomb atomic model is valid:

$$p_1 = n.R_i.\underline{T}_1.\rho \qquad p_2 = \rho.PE_1/Y \qquad p_3 = k_B.\underline{T}_1.\rho \qquad p_4 = kB.T_1.\rho/m_a \quad \{N/m^2\}$$

	Helium	Nitrogen	Oxygen	Fluorine	Neon	Chlorine	Argon
ρ	9.17599E-07	0.964387867	0.295545871	1.60E-05	1.60E-05	0.003214	0.016379
p_1	0.567390018	170406.8635	45716.82756	2.084366376	1.964042434	224.3696207	1014.760406
p_2	0.567390018	170406.8635	45716.82756	2.084366376	1.964042434	224.3696207	1014.760406
p_3	0.567390018	170406.8635	45716.82756	2.084366376	1.964042434	224.3696207	1014.760406
p_4	0.567390018	170406.8635	45716.82756	2.084366376	1.964042434	224.3696207	1014.760406

Note: the above values should be divided by the number of atoms (N) in the molecule.

8.4.3 Noble Gases

We can predict the noble gases by setting $\Gamma = 9.(\psi-1)$; they occur when 'Γ' is equal or close to an integer as can be seen in the plot below.

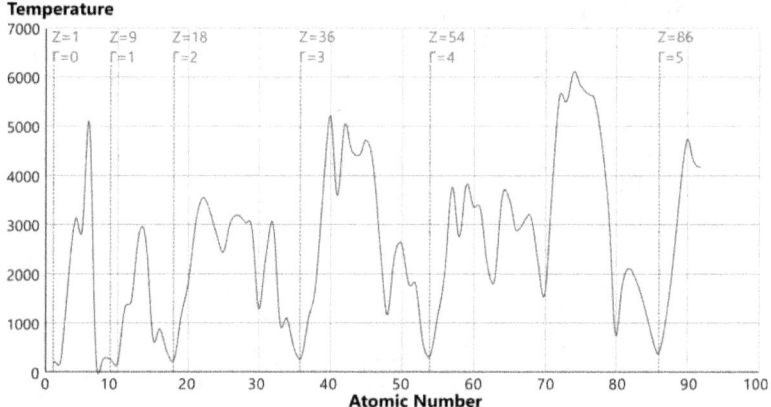

In fact, it is highly likely that all matter performance can be explained via viscosity and pressure using ψ, F_m & F_e.

8.5 Calculations

The density of viscous matter and its gas-transition temperature are related.

As can be seen in the figure below, an unmistakable relationship exists between the viscous density and gas-transition temperature of every element.

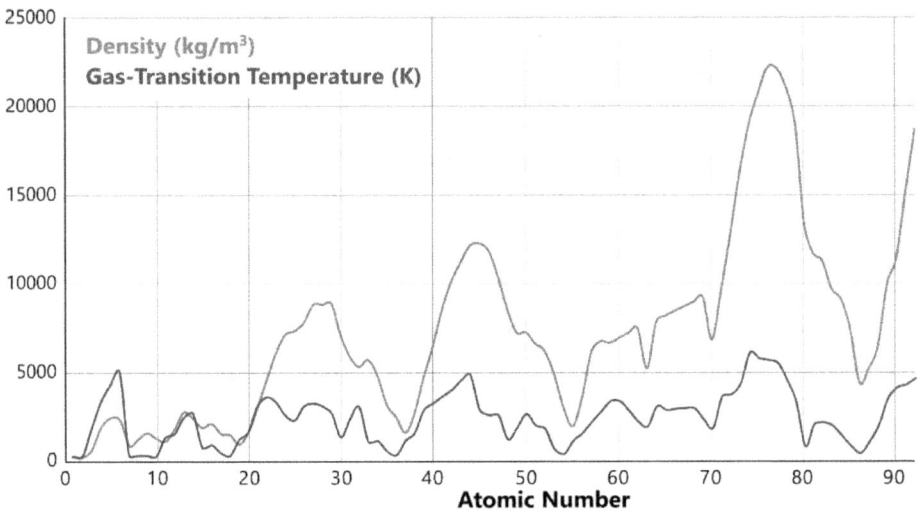

In fact, if you know the gas-transition temperature of elemental matter, you can calculate its viscous density, and vice versa. But as of today, I have not yet been able to predict either property simply from an element's atomic number (Z) and its relative atomic mass (RAM), which is my ultimate goal. However, until I achieve this goal, as the density of a pure crystal of elemental matter, which is not dependent upon temperature[#], atmospheric pressure or gravitational acceleration, I assume documented densities are correct, but not documented gas-transition temperatures.

[#] yes, density does vary with temperature, but only slightly, as phases change and the resultant inter-atomic force ($F = F_m - F_e$) weaken.

However, the inter-atomic forces between atoms (F_m & F_e) are predictable and between them capable of predicting all the other properties of the elements (chapters 8.3 and 8.4). These may be established as follows:

The Mathematical Laws of Natural Science

Mean inter-atomic spacing may be determined from an element's density thus;

element densities; $\rho = m_a/d^3$ {kg/m³}

and their mean inter-atomic spacing; $d = \sqrt[3]{[m_a/\rho]}$ {m}

The electrical force of repulsion between adjacent atoms due to the electrical charge in their nucleic protons, which varies linearly with temperature, may be calculated thus;

$$F_e = PE_1 / Y.d = m_e.\underline{T}/X / Y.d = k_B.\underline{T}/d \quad \{N\}$$

The magnetic force of attraction between adjacent atoms due to the magnetic field generated by their proton-electron pairs, which remains constant irrespective of temperature, may be calculated thus;

$$F_m = \mu_o.\xi_m/\zeta^3 . I_s^2 . (2\pi)^2 . (R_s/d)^3 . N.RAM.m_n/m_p \quad \{N\}$$

but is identical to the force predicted by Newton's constant of motion for the orbiting electron;

$$F_m = \rho.h_e^2 / \zeta^3 \quad \{N\}$$

As with the density and temperature (figure above), there is an unmistakable relationship between F_m and F_e.

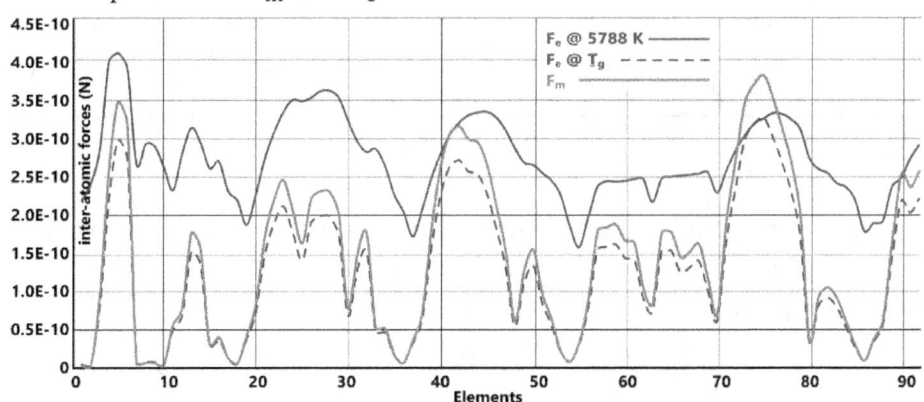

Inter-atomic pressure may be calculated from the electrical charge force thus:

$$p = F_e/d^2 + p_{atm} = k_B.\underline{T}/d^3 + p_{atm} = R_i.\underline{T}.\rho / (N.RAM/1000) + p_{atm} \quad \{N/m^2\}$$

where N = molecular number (e.g. 2 for N_2, O_2, etc.) and p_{atm} = atmospheric pressure

The Mathematical Laws of Natural Science

The only variable missing from these calculations is the lattice factor (ζ), which must be applied to F_m thus:

$$F_m = \rho \cdot h_e^2 / \zeta^3 \qquad \{N\}$$

This value should reflect the lattice structure of each element, which begins in - and is replicated from - its atomic nucleus.

As can be seen in the figure below, the lattice factor (ζ) of elemental matter appears not only to be related to the ratio of its density and its gas-transition temperature, but clearly identifies the noble gases (chapter 7.2.2). This plot also shows us that, whilst lattice structures are unique, they do follow a distinct pattern (Γ).

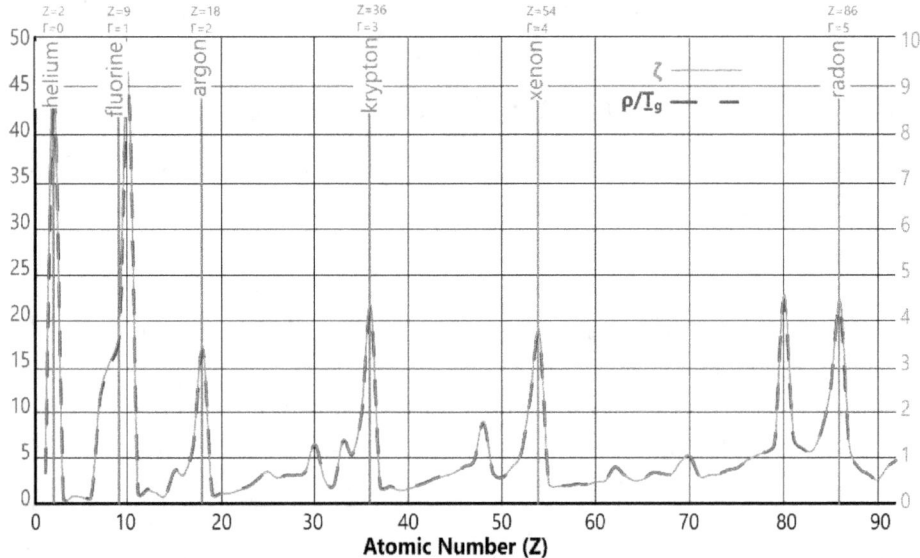

It is today believed that elemental matter in its liquid state does not occupy a lattice structure; this is incorrect. Whilst the lattice structure of any element may vary with a change in temperature, because it is defined by - and replicated from - its nucleic structure, a lattice structure will always exist at any temperature, and in any state; solid, liquid and gaseous. This lattice structure is the reason for Dalton's Law.

9 Orbits

Celestial orbits are the source of all universal energy.

Atomic orbits are responsible for all universal energy storage and transfer.

Orbits are ubiquitous; they are everywhere in the universe, both celestial and atomic. The energy in all these orbits originated from that released during the last 'Big-Bang'.

Orbital energy is constant and eternal (in a vacuum), and the source of internal body heat in celestial bodies through planetary spin, which is the source of neutron creation through atomic fission in bright stars and planets.

The heat we feel on the surface of our planet from stellar radiation is the EME generated by the proton-electron pairs in the sun's atoms.

There are no such things as binary stars. Those which today's astronomers claim are binary stars comprise one bright star, orbited by a bright planet. Like bright stars, bright planets have collected sufficient sub-satellite mass to create the heat required to generate fissionable energy within its core atoms.

A force-centre/satellite partnership will remain until the satellite is physically removed. Even satellite rubble left behind after impact will remain in the same orbit after destruction; which is the origin of asteroid belts.

The force-centre is always located at the focal point of the orbital system, which is on the major axis and the closest point to an orbiting satellite.

A satellite's orbital motion is maintained by the potential energy (PE) induced between it and its force-centre from its apogee, and subsequent deceleration induced by the same force after it passes its perigee. In a vacuum, this motion is perpetual; it will endure indefinitely.

Celestial orbital shape and energy (E) never change. The energy released during the last 'Big-Bang' was ≈7.3544E+60 Joules, which is the total energy (E) in all universal orbits.

Because orbital KE is positive and PE is negative, whilst a satellite's KE increases with its PE, total energy (E = KE + PE) remains constant. Also, because of the conservation of energy and the fact that the orbit is a symmetrical ellipse, the potential & kinetic energies on both sides of its orbital principal axis are equal and opposite.

The Mathematical Laws of Natural Science

9.1 Terminology

An **orbit** is the path followed by a satellite around its force-centre. For example, our moon is in orbit around our Earth (a planet), which is in orbit around our sun (a star), which is in orbit around Hades (a galactic force-centre). Electrons orbit their protons (a proton-electron pair).

An orbiting body (or *mass*) is referred to as a **satellite** and the body about which it orbits is referred to as its **force-centre.**

Solar Orbit: A star's orbital path around its galactic force-centre.

Planetary Orbit: A planet's orbital path around its star.

Lunar Orbit: A moon's orbital path around its planet.

Atomic Orbit: An electron's orbital path around its proton.

An **orbital system** is a force-centre with all of its satellites and sub-satellites, all of which have group names such as:

Collectively, everything (including the force-centre) orbiting a ...

... galactic force-centre is called a **galaxy**

... star is called a **solar system**

... planet is called a **lunar system**

... proton is called a **proton-electron pair**

An orbit is *always* a perfect ellipse, exactly as Johannes Kepler stated. An ellipse can be any two-dimensional (flat) elliptical shape including a circle.

There is a major difference between a genuine elliptical orbit, i.e. one of which its axes *are not* identical in length, and a circular orbit, i.e. one of which both its axes *are* identical in length, and a linear orbit:

Satellites following **elliptical orbits** (e.g. stars, planets, moons, comets, etc.) persist due to their satellite's kinetic energy and the potential energies (Newton's constant of motion) between their satellite and force-centre.

Satellites following **circular orbits** (e.g. electrons, geo-political, etc.) keep going because they provide their own kinetic energy.

Satellites following **linear orbits** (e.g. galactic force-centres) follow a single circuit. Subsequent orbits must be re-initiated; e.g. the '*Big-Bang*'.

The Mathematical Laws of Natural Science

Velocity in orbits refers only to the curvilinear motion of a satellite in its path around its force-centre. It does not refer to rotational (angular) motion in a satellite or its force-centre.

Mass is magnetic charge.

Gravity is the potential energy generated by magnetic charge.

Force is energy per unit distance.

Energy is a force applied over a specified distance.

Kinetic energy is the energy in a satellite due to its velocity.

Potential energy is the attractive/repulsive energy between a satellite and its force-centre.

Planetary Spin is the angular velocity (radians per second) in a body rotating about an axis that passes through its centre of *mass*.

9.2 Orbital Laws

The following fundamental laws apply to every natural orbit:

1) Every orbital system has only one force-centre and at least one satellite.

2) Altering the mass of a satellite will not affect its orbital shape or period.

3) Every orbital path is an <u>exact</u> ellipse, the eccentricity of which is; $0 \leq e \leq 1$

4) Every orbit about the same force-centre has an identical constant of proportionality (K).

5) Other than linear orbits; the constant of motion (h) of a satellite applies throughout its orbit.

6) A satellite's total energy (E=PE+KE) is a constant throughout its orbit.

7) Spin plays no part in the mathematical laws of an orbit.

8) Once united, an orbital partnership will remain intact until or unless it is physically broken.

9) In circular orbits (e = 0), a satellite must provide its own kinetic energy, which is always exactly half the potential energy between it and its force-centre;

$$PE = 2.KE = 2 \cdot \tfrac{1}{2}.m.v^2 = m.v^2 \qquad \{J\}$$

9.3 Centrifugal Force

Centrifugal force in a satellite is the potential force opposing its force-centre's gravitational force. A satellite will only remain in orbit as long as the centrifugal and gravitational forces are equal; *station keeping*.

For any given satellite, centrifugal force has one constant (mass) and one variable; (acceleration).

Whilst the calculation for centrifugal acceleration is relatively straightforward in circular orbits;

$F_c = m.v^2/R$ {N}

it isn't quite so straightforward for elliptical orbits.

9.3.1 Circular Orbits

Circular orbits are those with an eccentricity; e=0

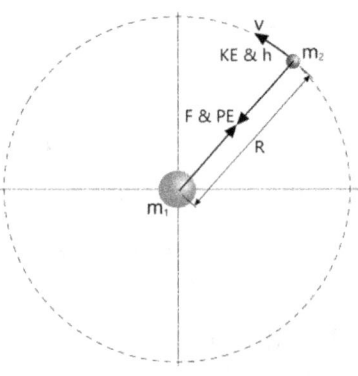

These orbits are not self-generating, in that the satellite must provide its own kinetic energy. They refer specifically to man-made geo-political satellites and proton-electron pairs, and are the justification for Henri Poincaré's formula; $E=mc^2$

Whilst these are calculated using exactly the same formulas as for elliptical orbits, their circular nature makes these calculations much simpler.

The Mathematical Laws of Natural Science

9.3.2 Elliptical Orbits

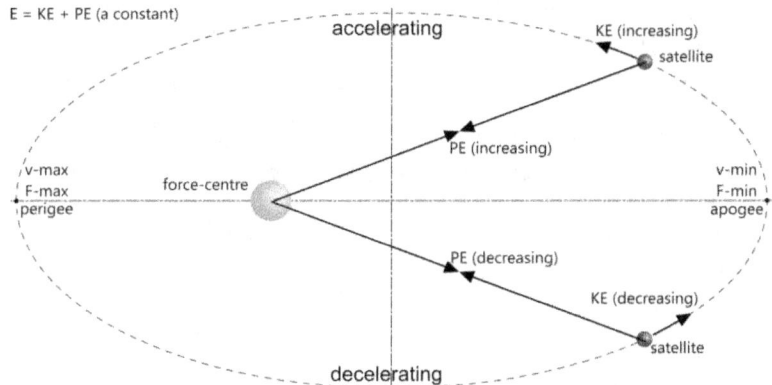

Elliptical orbits are those with an eccentricity; $0 < e < 1$

These orbits are self-generating in that the variable potential energy in a satellite induces in it a variable kinetic energy, which reach a maximum and minimum at its perigee and apogee respectively.

As such, this orbit will continue unchanging until the next 'Big-Bang', but only in a vacuum; if there are no atmospheric particles (e.g. dark matter) to absorb the satellite's kinetic energy.

In celestial bodies, the competing energies (potential and kinetic) in a force-centre, its satellites and sub-satellites are responsible for generating internal frictional heat within a satellite through planetary spin.

9.4 Linear Orbits

A linear orbit is the path of a satellite that departs its force-centre in a straight-line and continues travelling until the potential (gravitational) energy causes outward travel to cease, after which it will return to its force-centre following the same linear path.

This type of orbit is that which follows a 'Big-Bang' by the matter ejected from the ultimate body, in which the satellites are galactic force-centres. The time taken for all the galaxies to complete their linear orbits is a universal period.

9.5 Station Keeping

When a displacement force attracts a satellite, attempting to pull it off course, a restoring force keeps it in its orbital path, maintaining balance between its centrifugal and potential forces; i.e. which must be equal and opposite.

This relationship is what maintains a satellite's orbital path. Even if a satellite is destroyed through impact, the rubble will remain in the pre-impact orbit.

This process requires a perfect elliptical orbit; exactly as Kepler explained. It does not work with Relativity.

9.6 Orbital Planes

The spin induced in a force-centre by its orbiting satellites, influences all the orbital planes in an orbital system.

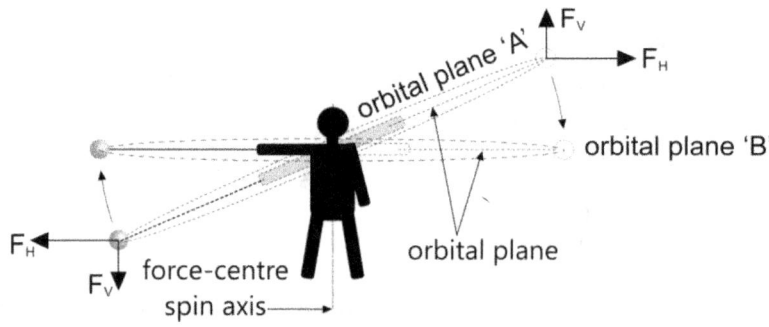

The rotational kinetic energy in a force-centre naturally causes the orbital planes of its satellites to settle at 90° to the force-centre's spin axis. This phenomenon can be demonstrated by attempting to swing a ball about *orbital plane 'A'* in which 'Fv' is non-zero, and subsequently through *orbital plane 'B'*, where 'Fv' is zero. The same forces are at work in celestial orbital systems; unless an external force is preventing natural settlement, the orbital plane will always settle at the lowest energy condition.

9.7 Orbital Precession

Because station-keeping maintains a satellite's orbital shape, the effect of gravitational attraction by external bodies is to induce a torque on a satellite's orbital axes, causing them to rotate about the orbital force-centre.

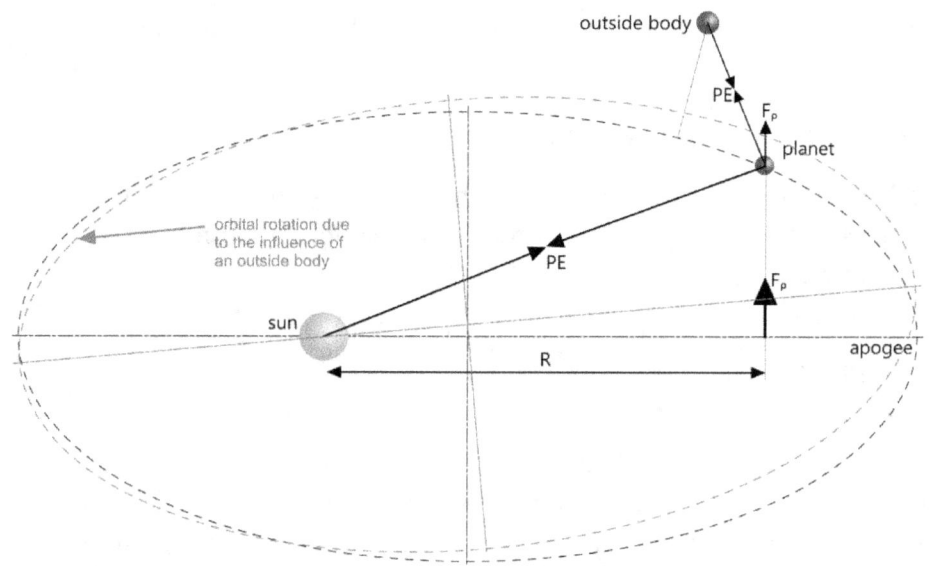

This orbital rotation is the reason celestial satellites in adjacent orbits occasionally impact, creating galactic and solar comets, which occasionally become trapped by other satellites; new planets and moons.

The Mathematical Laws of Natural Science

9.8 Calculations

The mathematical laws for orbital systems were given to us by Copernicus, Kepler, Galileo and Isaac Newton before the end of the 17th century. They remain valid today, irrespective of the orbital system variables.

Refer to Appendix A-5 for an analysis of these mathematical laws.

9.8.1   Newton's Laws of Orbital Motion

The laws of orbital motion are the mathematical formulas and statements that describe the properties of a satellite's curvilinear motion around its force-centre.

Newton's laws of orbital motion come in two parts:
1) orbital shape and period
2) satellite performance

The calculation results from one cannot be used to validate the other because the force-centre *alone* defines orbital shape and period. I.e. it is possible (in theory and in practice) to swap any satellite from its own orbit with that of another without altering the orbital periods or shapes. The only differences will be the satellite's performance.

In other words; whilst you can calculate the mass of a force-centre from the dimensional shape of any one of its satellite orbits, you cannot calculate satellite mass in the same way.

A satellite's orbit (shape and period) is not altered by its own orbiting secondary satellites (e.g. moons).

All orbits are elliptical, the eccentricity of which must be; '$0 \leq e < 1$'

The following important mathematical laws apply to all orbits, irrespective of eccentricity.

The key formula for orbital motion originally defined by William Gilbert, and later described as follows by Isaac Newton:

   Potential Force: $F = G \cdot m_1 \cdot m_2 / R^2$ {N}

which can be modified to define the following:

   Potential Energy: $PE = G \cdot m_1 \cdot m_2 / R$ {J}
   Potential Acceleration: $g = G \cdot m_1 / R^2$ {m/s^2}

The Mathematical Laws of Natural Science

Alternative formulas for satellite velocity and potential acceleration:

$v = h/R$ \{m/s\}

$g = F/m_2$ \{m/s^2\}

Throughout any orbit, E is <u>always</u> constant, whilst PE and KE both vary, which is only possible because PE is negative and KE is positive:

$E = -PE + KE$ \{N\}

Constant of Proportionality:

$K = t^2/a^3 = (2.\pi)^2 / G.m_1$ \{s^2/m^3\}

Where:
G; is Newton's gravitational constant
h; is Newton's constant of motion
m_1; represents the mass of the force-centre
m_2; represents the mass of the satellite
R; represents the distance between the centres of m_1 and m_2
F; represents the potential force acting between m_1 and m_2
PE; represents the potential energy acting between m_1 and m_2
t; represents the total orbital period
a; represents half the length of the orbital major axis

A useful tip from Kepler that was later verified by Newton is that the relationship between the swept area inside the ellipse for any given satellite's orbital period will always be identical.

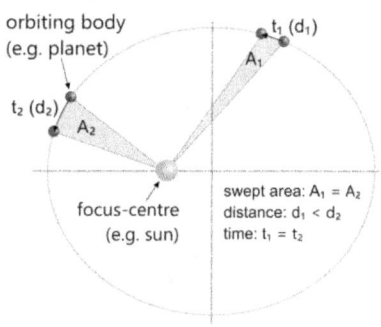

The calculations for all elliptical orbits are provided below (chapter 9.8.3). They have been derived from Isaac Newton's Philosophiæ Naturalis Principia Mathematica, and They work perfectly for all elliptical orbits irrespective of satellite performance.

9.8.2 Centrifugal Force

Circular Orbits:

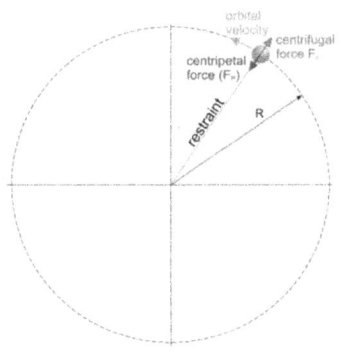

If you swing a ball - tied to a length of string - around your head, *centrifugal* force is pushing the ball away from you. But it also induces a tensile force in the string, pulling the ball towards you (*centripetal* force), *potential force*; the equivalent of *gravity*.

Christiaan Huygens gave us the mathematical relationship between this and its velocity in a circular orbit;

$a = v^2/R$ {m/s^2}

where 'v' is its curvilinear velocity and 'R' is its orbital radius.

The Mathematical Laws of Natural Science

Elliptical Orbits:

However, for elliptical orbits, the above orbital velocity (v) must be modified (v_c) to provide the correct gravitational acceleration, which would otherwise look like this:

$$a = v.v^A / R.(1-e)$$

where 'v^A' is the orbital velocity at the orbital apogee, 'e' is the orbital eccentricity and R is the orbital radius (at any orbital angle 'θ').

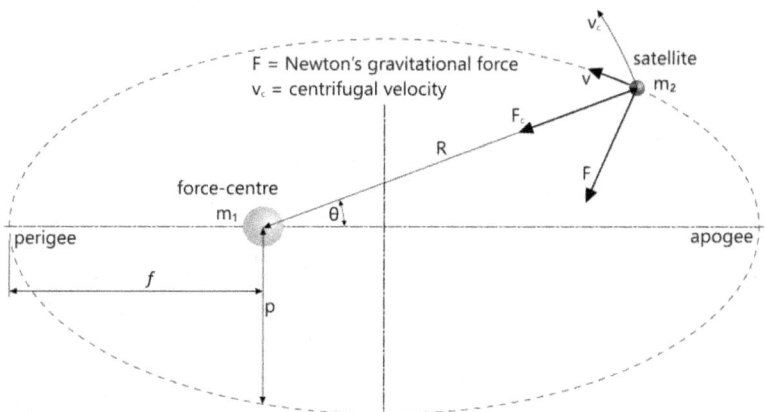

Centrifugal force in elliptical orbits may be calculated at any orbital angle (θ) and orbital radius (R) thus:

$$\alpha = \sqrt{[^4/_3.\pi]}$$

$$\kappa = \sqrt{[\,(f.\sin(\theta/2)^\alpha + p.\cos(\theta/2)^\alpha) / (f.\cos(\theta/2)^\alpha + p.\sin(\theta/2)^\alpha)\,]}$$

$$v_e = \kappa.v \qquad \{m/s\}$$

$$F_e = m_2.v_e^2/R \qquad \{N\}$$

which may be simplified at the orbital extremes as follows:

@ the perigee of an ellipse; $F_c = F . f/p = F / (1+e)$ {N}

@ the apogee of an ellipse; $F_c = F . p/f = F . (1+e)$ {N}

The Mathematical Laws of Natural Science

9.8.3 Orbits

Refer to chapter 6.1 for the formulas relating to circular orbits. Those relating to elliptical orbits are provided below.

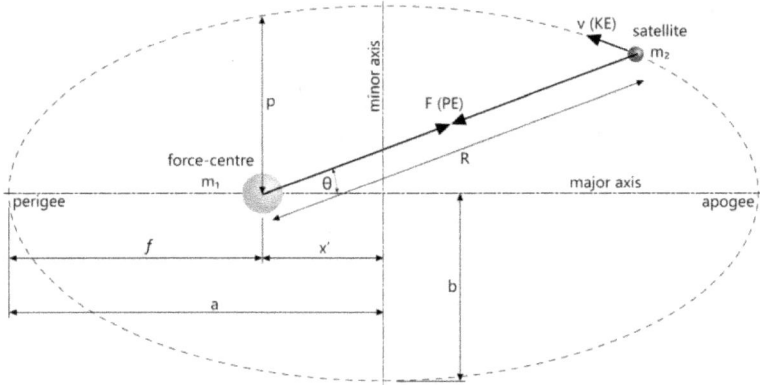

This Table is a list of the formulas that fully describe a satellite orbit.

Sym	Description	units
R^P	radius at orbital perigee ($R^P = 2.a - R^A$) [1]	kg
R^A	radius at orbital apogee ($R^A = 2.a - R^P$) [1]	kg
θ	angle through orbit [2]	c
t	orbital period	K

Table 9.1: *Input Data*
1) Once K has been established for the first orbit, only one of these values is required for all other satellites orbiting the same force-centre
2) $\theta = 0$ when satellite is at its apogee

Sym	Formula	Description	units
a	$(R^A + R^P)/2$ or $\sqrt[3]{t^2/K}$ [1]	major semi-axis	m
K	t^2/a^3	constant of proportionality [2]	s^2/m^3
f	R^P	orbital distance @ perigee	m
x'	a-f	orbit centre distance	m
e	$(-R^P + \sqrt{R^{P2} - 4.a.(R^P - a)})/2.a$	eccentricity	
b	$\sqrt{a^2.(1-e^2)}$	minor semi-axis	m
p	$a.(1-e^2)$	half-parameter	m
A	$\pi.a.b$	orbit total area	m^2
L	$\pi . \sqrt{2.(a^2+b^2) - (a-b)^2 / 2.2}$	orbital circumference	m
R	$p/[1 - e.\cos(\theta)]$	orbital radius @ θ	m
v	$2.A / t.R$	orbital velocity @ R	m/s
g	$-v.v^A / R.(1-e)$ [3]	gravitational acceleration	m/s^2

Table 9.2: *Orbital Shape*
1) This formula may be used to calculate a once K has been established for the first orbit
2) This value may be used to calculate R or R for all other orbits around the same force-centre
3) v^A = satellite velocity at its orbital apogee ($2.A / t.R^A$)

The Mathematical Laws of Natural Science

Sym	Formula	Description	units
m_1	$(2.\pi)^2 / G.K$	force-centre mass [1]	kg
m_2	input	satellite mass	kg

Table 9.3: *Body Mass*
1) Only one of these values is required for all orbits around the same force-centre once K has been established for the first orbit
2) Only required for a satellite between its orbital apogee and perigee

Sym	Formula	Description	units
F	$g.m_2$	potential force on satellite	N
PE	F.R	potential energy on satellite	J
KE	$\frac{1}{2}.m_2.v^2$	kinetic energy in satellite [1]	J
E	PE+KE	total energy [2]	J
p	$m_2.v$	satellite momentum	kg.m/s
h	v.R	Newton's motion constant	m²/s

Table 9.4: *Satellite Performance*
1) Does not include angular momentum, which plays no part on Newton's laws of orbital motion
2) This value does not change throughout a satellite's orbit (it is a constant for the satellite in its orbit)

The mass of a satellite (m_2) can be established if you know the potential acceleration (g_s) at a specified radius (r_s); e.g. at its surface:

either by using Isaac Newton's gravitational constant; $m_2 = -g_s.r_s^2 / G$

or from the following formula if you've forgotten 'G'; $m_2 = m_1 . g_s/g^P . (r_s/R^P)^2$

The relationship between K, h & satellite & force-centre masses (m_2, m_1) is;

$K = t^2/a^3 = (2\pi)^2 / G.m_1 = (2\pi)^2 / v^P.v^A.a$ {s²/m³}

$h = R.v$ {m²/s}

$m_2.v^P.v^A / E = -2$

$E.(K.h)^2/m_2 = 8.\pi^4$

$2.E/m_2 . (h / G.m_1)^2 = 1$

$G . \sqrt{[m_2 / 2.E.(v^P.v^A)]} = R / m_1$ {m/kg}

where v^P & v^A are satellite velocity at the orbital perigee and apogee respectively

The Mathematical Laws of Natural Science

9.8.4 Linear Orbits

All the laws of motion that apply to elliptical orbits also apply to linear orbits, except for Isaac Newton's constant of motion, which is a variable in the case of linear orbits.

You may calculate the properties of the orbit at any time during outward travel thus,

Governing Formulas:	Intermittent Values:
$u = \sqrt{[2.E/(m^u - m_1)]}$	$v = u + a.t$
$K = (2\pi)^2 / (m_1.G)$	$K = (2\pi)^2 . KE/PE . 2/(R.v^2)$
$R_o = \frac{1}{2}.u.t_o$	$R = u.t + \frac{1}{2}.a.t^2$
$t_o = (4\pi)^2 / K.u^3$	$KE = \frac{1}{2}.m_2.v^2$
$E_2 = E . m_2/m^u$	$PE = G.m_1.m_2/R$
$a = -u^2 / 2.R_o$	$g = -G.m_1/R^2$
where; m^u = *ultimate body mass*	

after which, the satellite will return to its force-centre using exactly the same formulas, but altering deceleration (-a) to acceleration (+a).

At the start of orbit: $KE = E$, $PE = 0$ & $g = 0$

At the point where outward travel ceases: $KE = 0$, $PE = KE$ & $g = a$

9.8.5 Station-Keeping

This is a graphical representation of the restoration force on the earth $45°$ through its orbital path from its apogee.

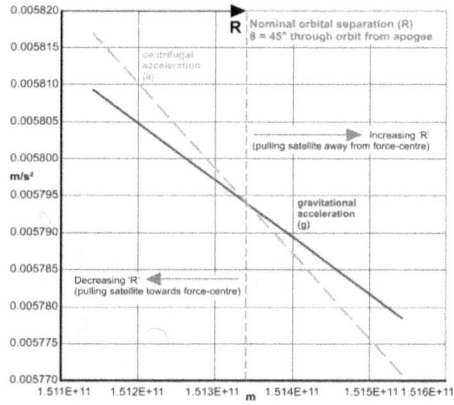

As the earth is pulled away from the sun (increasing R), gravitational acceleration (g) increases faster than centrifugal acceleration (a), pulling the earth back towards the orbital path when the displacement force is released.

As Earth is pulled towards the sun (decreasing R), centrifugal acceleration (a) increases faster than gravitational acceleration (g), pulling the earth back towards the orbital path when the displacement force is released.

As you can see, exact balance occurs at the orbital path: at nominal orbital separation 'R'.

The following graph shows the discrepancy between the velocity ratio calculated using NASA's orbital dimensions (*empirical*) and those using the elliptical centrifugal force (*calculated*) for the earth throughout a complete orbit (θ). The discrepancy is due to periodic variations in our sun's orbit, mistaken observation and the calculated version being theoretically perfect.

This plot was calculated using the same formulas and input data as for the plot at the top of this page for station-keeping.

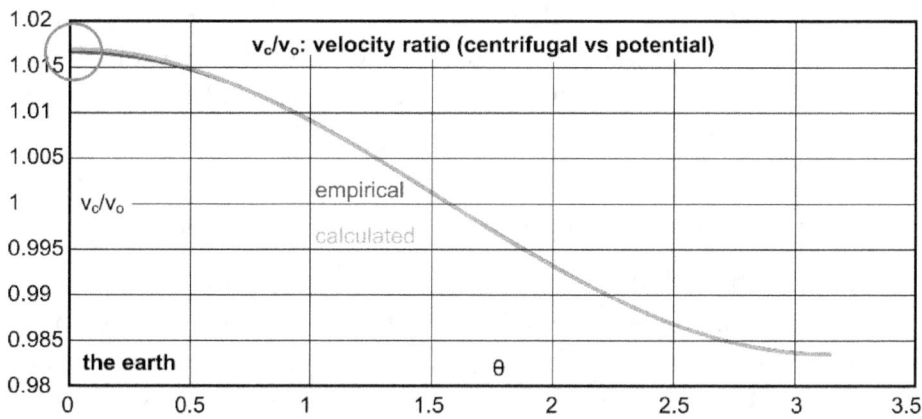

The Mathematical Laws of Natural Science

All the orbits in the solar system have been plotted and that for the earth shows minimal discrepancy because we can measure its orbital dimensions with reasonable accuracy. Jupiter, however, shows a larger discrepancy (see below) as we are less able to define its orbital properties accurately.

However, we now have the opportunity to calculate accurate orbits for all of our solar system satellites.

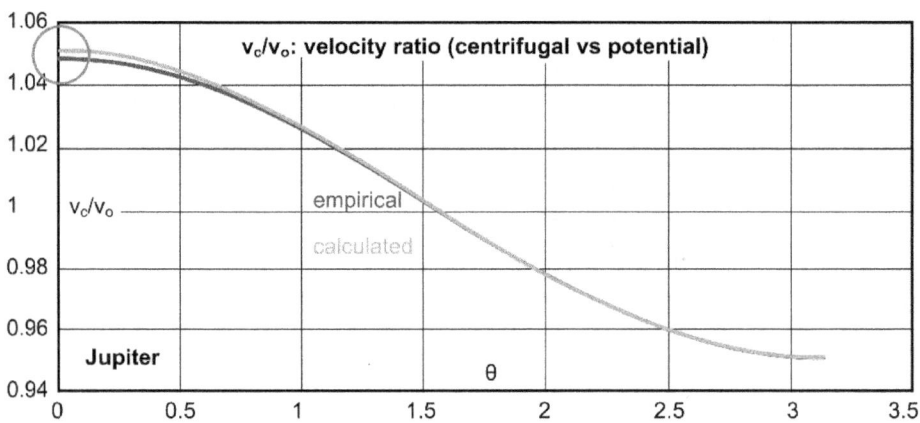

9.8.6 Orbital Precession

It is currently believed that satellites are temporarily influenced by other celestial bodies. The potential energy between the satellite and the external body will pull them together, out of their respective orbits temporarily altering their orbital paths. The satellite will accelerate as it travels towards the external body and decelerate (relative to the external influence) after passing it. However, these variations are effectively cancelled out as a result of Kepler's and Newton's 'equal time swept-area' law, station-keeping and the conservation of energy.

This is, however, incorrect. What actually happens is; whilst the orbital period is maintained, these temporary influences apply a torque ($T_p = F_p \times R$) to the orbital axes causing a gradual rotation (of the orbit) about its force-centre. Their frequency and magnitude define the rate of orbital precession. It is this orbital rotation that is occasionally responsible for the impact between adjacent satellites and the creation of comets.

9.8.7 Orbital Planes

Natural orbits obey a fundamental law in that they <u>always</u> settle at their lowest energy condition.

This condition means that only longitudinal (radial) forces are holding a satellite in orbit; i.e. no lateral forces are present. Out-of-plane orbits occur where lateral forces are induced in the satellite by adjacent force-centres and satellites.

All orbital planes about a single force-centre coincide because this is the lowest energy condition in each orbit:

$$F_h < \sqrt{[F_h^2 + F_v^2]}$$

F_h represents the potential force in orbit @ 90° to the force-centre's spin axis.

F_v represents an out-of-plane force when ≠ 90° to the force-centre's spin axis.

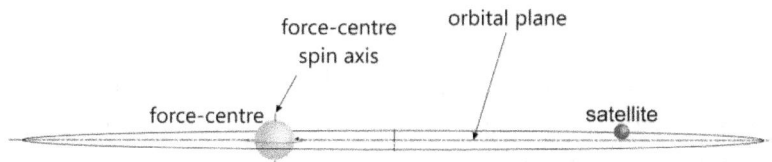

The satellite with the greatest kinetic and potential energy will dominate the force-centre's spin axis, after which, all lesser orbits about the same force-centre will eventually settle into the same plane; 90° to the force-centre's spin axis

Out-of-plane orbits are either newly trapped satellites that have yet to settle, or are subject to external forces preventing natural settlement.

The Mathematical Laws of Natural Science

9.8.8 Planetary Mass

We can calculate the mass of a celestial body in a number of ways, dependent upon your circumstances.

Beneath Your Feet

If you are standing on a planet, the radius of which you know, you can discover its mass simply by dropping any mass from any height in a vacuum. The acceleration of the mass can be established like this:

> Drop height: $s = u.t + \frac{1}{2}.a.t^2$ {m}
> because; $u = 0$: $a = 2.s/t^2$ {m/s}
> its mass is established using Isaac Newton's force formula thus:
> $m = a.R^2/G$ {kg}

Force-Centre

If the planet is a force-centre, i.e. it has satellites of its own, its mass may be found from its satellite orbits using Isaac Newton's constant of proportionality (K) like this:

> $m = (2\pi)^2 / G.K$ {kg}

Note: if this calculation does not provide the answer you are expecting, your satellite's orbital properties (a or t) are incorrect, where; 'a' is the satellite's half-major-axis, and 't' is its orbital period.

Lone Planet

If your planet is all alone; i.e. it has no satellites of its own, you will not be able to calculate its mass from its orbital dimensions or performance. This is why:

> $E_1 = \delta KE . (r/R)^2 = \frac{1}{2}.m.(v^{P2} - v^{A2}) . (r/R)^2$ {J}
> $E_1 = \frac{1}{2} . J . (\omega.|\omega|) = \frac{1}{2} . \frac{2}{5}.m.r^2 . (\omega.|\omega|)$ {J}
> Therefore ...
> $\frac{1}{2}.\cancel{m}.(v^{P2} - v^{A2}) . (r/R)^2 = \frac{1}{2} . \frac{2}{5}.\cancel{m}.r^2 . (\omega.|\omega|)$
> ... the mass components cancel out.

This is why you can swap planets from one orbit to another without affecting their orbital dimensions.

10 Spin Theory

Spin is the angular motion (rotation) of a body, and the source of all universal electro-magnetic energy through internal friction.

It can be predicted with the same degree of accuracy as Newton's laws of orbital motion from the potential and kinetic energies calculated using them. These same theories (formulas) apply to all spinning bodies, from Quanta to galactic force-centres

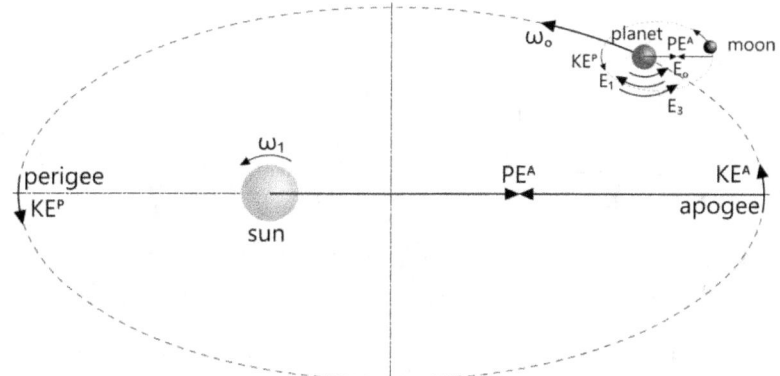

E_0: Potential energy between a force-centre and that of its satellite will naturally cause the satellite to spin at the same angular velocity and in the same rotational direction (prograde) as its orbit ($-\omega_0$).

E_1: A force-centre's own rotational kinetic energy induces spin in a satellite causing it to rotate in the opposite (retrograde) direction (ω_0).

E_3: If a sub-satellite (e.g. moon) is orbiting in the same rotational direction as the satellite's orbit (prograde), the sub-satellite's kinetic and potential energies will cause the satellite to rotate also in a prograde direction ($-\omega_0$).

E_2: The angular kinetic energy in a satellite is the sum of the above:

$$E_2 = E_1 - E_0 - E_3$$

Only the potential and kinetic energies in a force centre, its orbiting satellite(s) and their sub-satellites induce spin in each other. It is not as difficult or complex as everybody appears to believe.

For example: The reason Venus spins in the opposite direction is;

E_3 (which is generated by a sub-satellite), is generally much greater than E_1 and E_0 and therefore defines a planet's rotational direction. As E_3 in a planet with no moon is zero, its orbital energies (E_1 & E_0) determine its spin direction. Being a (relatively) large planet, E_1 is the dominant factor in Venus' spin direction ($E_1 > E_0$).

The same argument applies to Mercury except in its case its smaller size means that E_0 is dominant ($E_0 > E_1$) so it spins in the same direction as the other planets in our solar system.

10.1   Chicken & Egg

Which came first; spin or orbit?

What you see in most films and documentaries is that the sun starts spinning and the planets follow it around. This is of course 'back-to-front'.

In order to generate spin, you need an appropriate energy. Spin theory teaches us that if a sun, planet or moon sat alone in space it would not spin.

Spin in our sun was first induced by the rotational energy in its force-centre (Hades) and it would have continued to spin at this rate had it not acquired its satellites (planets). However, our sun actually rotates at considerably more than this speed.

If angular kinetic energy in a force-centre induced orbital kinetic energy in its satellites, this transfer of energy would slow down the force-centre's rotation, which is obviously not the case. I.e. kinetic energy in the planets can and does induce rotational kinetic energy in the sun.

Therefore, the planets must have been orbiting long before the sun achieved rotation anywhere near its current rate.

The same argument applies to a spiral galaxy. Our sun got its initial spin (<2E-07 radians per second) from the spin energy in Hades, but Hades has no force-centre. Therefore, all of Hades spin comes from its orbiting satellites.

So, orbits came first!

The Mathematical Laws of Natural Science

10.2 No Moon

If the earth lost its moon, it would also lose its tilt (and its seasons), there would be insufficient internal heat energy to drive its continental plates and its magnetic field would vanish. The loss of internal friction would cause the earth's surface temperature to fall by about 210K.

The earth would therefore behave similarly to Venus except for its atmosphere, most of which would liquefy/solidify because the earth' surface receives only 52% of the sun's radiated heat compared to Venus.

Without its moon, one earth day would be 12450.152 hours (<519 current earth days) and the sun would rise in the West and set in the East just like Venus. There would be no seasons because the earth would lose its tilt.

Without a moon, the earth's 'E_3' would be zero and its rotational energy would become:

$E_2 = E_1 - E_0 - 0 = 3.207996E+23 - 2.144732E+23 = 1.063264E+23$ J

$E_2 = \frac{1}{2}J.\omega_0^2 \rightarrow \omega_0 = \sqrt{[\,2.E_2/J\,]} = 1.401854E-07$ c/s

$t_s = 2\pi/\omega_0 = 4.482055E+07$ s

or 12450.1516 hours (<519 days).

Moreover, the earth would lose 99.99963% of its internal [heat] energy; i.e. its surface temperature would fall by approximately 210 K: no more continental drift.

We can therefore conclude that the presence of our moon is essential for life on earth.

10.3 Polar Moment of Inertia

Polar moment of inertia (J) is a body's resistance to spin and calculated as follows:

$$J = \tfrac{2}{5}.m.r^2 \qquad \{kg.m^2\}$$

where:
'm' is the body's mass
'r' is its outside radius

This formula only works as shown, however, if the body comprises a *mass* of constant density, which is not the case for celestial bodies, such as; planets, stars, moons, etc. *Gravity* tends to ensure that the denser matter migrates towards their cores.

This problem can be solved by incorporating a radial modifier (Δ) in the above formula thus ...

$$J = \tfrac{2}{5}.m.(\Delta.r)^2 \qquad \{kg.m^2\}$$

... which defines a body's radial centre of *mass* according to its variable density.

If you know a body's angular velocity (ω) you can calculate its radial modifier (Δ). Alternatively, if you know its radial modifier, you can calculate its angular velocity:

$$E = \tfrac{1}{2}.J.\omega.|\omega| \qquad \{J\}$$

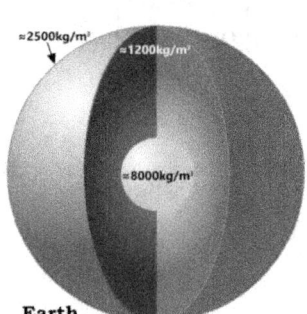

Earth

The greater the 'Δ' value, the lower the density variation;

Δ < 1 means the core density is greater than its surface density (normal situation)

Δ = 1 means that the entire body is homogeneous

Δ > 1 indicates that the body is being pulled into a local orbit

'Δ' can help us to determine a satellite's internal structure when used in conjunction with its core pressure.

10.4 Internal Heat

Potential and kinetic energies act differently on orbiting satellites. PE acts at their cores and KE acts throughout their *mass*. This conflict is what causes a satellite's mantle to spin relative to its core.

Core-mantle rotation ($\delta\omega$) is the term used here for this relative spin-rate, and E_δ is the energy generating $\delta\omega$.

The potential energy between a force-centre and its satellite (E_0) tries to hold a satellite's core at an angular velocity commensurate with the satellite's orbital period; ω_0.

The kinetic energies induced in a satellite (E_1 & E_3) act throughout a satellite's *mass* and are trying to overcome ω_0

This rotational conflict causes internal friction between a satellite's core and its mantle, generating internal heat. The greater a satellite's sub-satellite *mass*-population (E_3), the greater the satellite's internal heat.

This conflict is also responsible for generating a satellite's magnetic field, between its core and its mantle. The greater a satellite's sub-satellite *mass*-population, the greater will be the satellite's magnetic field.

Satellites with no sub-satellite(s) ($E_3 = 0$) will generate very little internal heat; i.e. they will be relatively cold, nor will they generate a magnetic field.

Iron is by far the most common element in the universe, and therefore in all universal bodies. It is therefore almost guaranteed that the core of all substantial celestial bodies (stars and planets) with sub-satellites (planets and moons) will comprise almost 100% iron, the density of which will be ≈ 7870 kg/m^3.

The Mathematical Laws of Natural Science

10.5 Magnetic Field

A satellite's polar magnetic field is generated by the relative rotation of the same electro-magnetic charges in the same Quanta as in the proton-electron pair; but in a celestial body, the Quanta involved are considerably more numerous.

The fact that only non-hollow planets (e.g. excluding Mars) with moons, actually generate a magnetic field normal to the orbital plane of their satellites, and those that generate the greatest spin-energy (E_δ) also generate the greatest magnetic field is additional proof of spin theory.

As I have set prograde direction as negative (E_3 is negative for the earth), and because '$\delta\omega$' is also negative, the earth's mantle (including its crust) must be spinning in a prograde direction. Using the right-hand rule, the magnetic

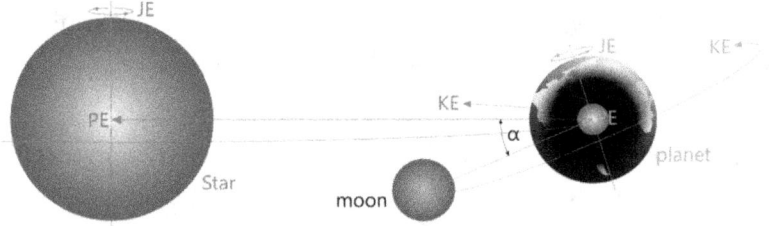

north pole of the earth should be pointing in the direction of our North Pole; which it is.

The earth's lunar tilt angle ($\alpha = 23.4°$) is a clear indication that the earth acquired its moon from outside its solar system (galactic comets), as is the case for all of the moons in our solar system, and probably, most of its planets.

If a planet's lunar orbital plane is not coincident with its solar orbital plane, its mantle and core will spin on different axes, generating an angular difference between its true (physical) North and its magnetic North.

The Mathematical Laws of Natural Science

10.5.1 Magnetic Reversal

A worrying aspect of the earth's magnetic field is that the relative spin induced in the earth and its core by our moon and our sun will not reverse unless either the earth or its moon changes orbital direction, which is highly unlikely. Something external to the earth (extra-terrestrial) must therefore cause this reversal to occur periodically.

It would appear that the earth's magnetic reversal can only be explained by flipping it through 180°; switching north and south poles!

We already know that our solar system has a number of orbiting comets, so it is highly likely that the Milky Way also has its own 'comets', and these could be planet sized. Therefore, a large galactic comet may well be responsible for flipping the earth and/or any other planet in the solar system as it passes close by.

10.6 Goodricke & Algol

In 1784, John Goodricke discovered the [supposed] binary nature of the star originally named Algol. As one of the binary stars passed in front of the other, their combined brightness dimmed, revealing two important facts:

1) One of the stars is bright (hot) and the other is dark (cold),

2) Only the bright star was a force-centre for an orbital system.

The heat generated by the *bright* star is due to its dedicated satellite population. The other darker star (or large planet) is actually in orbit around the bright-star but has few satellites of its own. In which case, the dark partner can generate no internal heat.

Once a star (force-centre) has acquired its satellites, it can trap a twin but it will not share its satellites.

This discovery also reinforces the facts that stars generate their own internal heat from their satellite population and the proton-electron pair behaviour in atoms.

The Mathematical Laws of Natural Science

10.7 Calculations

There are two calculation methods that can be used here, dependent upon whether or not a satellite's radial modifier (Δ) is known. In both calculation procedures you need the following information from the satellite's orbital calculations:

KE^A: kinetic energy of the satellite at its orbital apogee
KE^P: kinetic energy of the satellite at its orbital perigee
PE^A: potential energy of the satellite at its orbital apogee
a: half the length of the major axis of the orbital ellipse
r: satellite's radius
θ: angle of inclination of the satellite's orbital plane relative to its force-centre's spin-axis. '$\theta = 0$' when the orbital plane is 90° to the spin-axis

Note: |?| is the modulus or positive value; '?'
sign(?) refers to the polarity of value '?' (+1 or -1)

The results from these calculations can be used to determine relative core-mantle rotation and core pressures.

10.7.1 Radial modifier (Δ) known

If a satellite's radial modifier is known, the following procedure will establish its component angular velocities:

$$\omega_0 = 2\pi / t_0 \qquad \{^c/s\}$$
where: 't_0' is the satellite's orbital period

$$E_0 = \tfrac{1}{2} J . \omega_0 | \omega_0 | \qquad \{J\}$$

$$E_1 = \delta KE . (r/a)^2 \qquad \{J\}$$
where: $\delta KE = KE^P - KE^A$ and KE^P & KE^A are for the satellite

$$\omega_1 = \text{sign}(E_1) . \sqrt{[\, 2.|E_1| / J\,]} \qquad \{^c/s\}$$

$$E_2 = E_1 - E_0 - E_3 \qquad \{J\}$$
$$\omega_2 = \text{sign}(E_2) . \sqrt{[\, 2.|E_2| / J\,]} \qquad \{^c/s\}$$

$$E_3 = \text{sign}(\cos(\theta)) . (\Sigma KE^P + \Sigma PE^A) \qquad \{J\}$$
$$\omega_3 = \text{sign}(E_3) . \sqrt{[\, 2.|E_3| / J\,]} \qquad \{^c/s\}$$

Notes:
KE^P & PE^A are for the sub-satellite(s)
PE is always negative
'KE^P' is positive if the sub-satellite is orbiting in the same direction as the satellite and negative if it is orbiting in the opposite direction.
'θ' is -1 if the planet is tilted more than 90° and 1 is it tilted less than 90° relative to the plane of its sub-satellite

The Mathematical Laws of Natural Science

10.7.2 Radial modifier (Δ) unknown

If a satellite's orbital and spin periods are known, the component angular velocities can be established:

$E_1 = \delta KE \cdot (r/a)^2$ \{J\}

where: $\delta KE = KE^P - KE^A$ and KE^P & PE^A are for the satellite
'r' is the radius of the satellite & 'a' is half the length of the satellite's orbital major axis

$E_3 = \text{sign}(\cos(\theta)) \cdot (\Sigma KE^P + \Sigma PE^A)$ \{J\}

where: PE is always negative and KE^P & PE^A are for the sub-satellite(s)
'sign(cos(θ))' is either '1' or '-1' and reflects the tilt angle of the sub-satellite's orbital plane relative to the satellite's orbital plane

$\omega_0 = 2\pi / t_0$ \{^{c}/s\}

where: 't_0' is the satellite's orbital period

$\omega_2 = 2\pi / t_s$ \{^{c}/s\}

where: 't_s' is the satellite's spin period

The satellite's polar moment of inertia:

$J = 2 \cdot (E_1 - E_3) / (\omega_0 \cdot |\omega_0| + \omega_2 \cdot |\omega_2|)$ \{kg.m^2\}

From: $E_2 + E_0 = \frac{1}{2} J \omega_2^2 + \frac{1}{2} J \omega_0^2 = \frac{1}{2} J (\omega_2^2 + \omega_0^2) = E_1 - E_3$ & $\omega^2 = \omega \cdot |\omega|$

The satellite's radial modifier:

$\Delta = \sqrt{[5J / 2 \cdot m]} / r$
where: 'm' is the satellite's mass

$E_0 = \frac{1}{2} J \cdot \omega_0 \cdot |\omega_0|$ \{J\}

$E_2 = \frac{1}{2} J \cdot \omega_2 \cdot |\omega_2|$ \{J\}

$\omega_1 = \text{sign}(E_1) \cdot \sqrt{[2 \cdot |E_1| / J]}$ \{^{c}/s\}

$\omega_3 = \text{sign}(E_3) \cdot \sqrt{[2 \cdot |E_3| / J]}$ \{^{c}/s\}

10.7.3 Core Heat

The following constitutes the calculation sequence to establish the relative rotation rate of a planet's (or star's) core and its mantle:

Input Data

$\rho_c = 7870$ {kg/m³} (iron core)

$r_c = 1215000$ {m} (e.g. Earth)

$\rho_m = 5506.351327$ {kg/m³} (e.g. Earth)

$r_m = 6371000.685$ {m} (e.g. Earth)

$E_c = E_0$ {J}

$\Delta_c = 1$ (solid core of constant density)

$m_c = \rho_c / (^4/_3 . \pi . r_c^3)$ {kg}

$J_c = ^2/_5 . m_c . r_c^2$ {kg.m²}

$\omega_c = \sqrt{[\, 2.E_c / J_c \,]}$ {c/s}

Output Data

Core:

Mantle:

$m_m = \rho_m / (^4/_3 . \pi . r_m^3)$ {kg}

$J_m = J - J_c$ {kg.m²}

$\Delta_m = \sqrt{[\, ^5/_2 J_m / m_m . (r_m^2 - r_c^2) \,]}$

$E_m = E_3 - E_1$ {J}

$\omega_m = \sqrt{[\, 2.E_m / J_m \,]}$ {c/s}

Effective:

$\delta\omega = \omega_m - \omega_c$ {c/s}

$\delta E = E_m - E_c$ {J}

10.7.4 Magnetic Field

There are two competing energies driving the spin in the earth's core and its mantle:

-E_0 (-2.1447324E+23 J) is the sun's energy driving the core

E_1-E_3 (2.87704E+28 J) is the moon's energy driving the mantle (and core)

The polar moments of inertia:

Core: J = 3.49144112166E+34 kg.m²

Mantle: J_m = 1.07860404E+37 kg.m²

The angular velocities:

Core: ω = Sign(E_0) . $\sqrt{[2.|E_0|/J]}$ = -3.50509019131E-06 c/s

Mantle: ω_m = Sign(E_1-E_3) . $\sqrt{[2.|E_1-E_3|/J_m]}$ = 7.3039350764E-05 c/s

The differential angular velocity:

$\delta\omega$ = ω + ω_m = 6.95342605725E-05 c/s

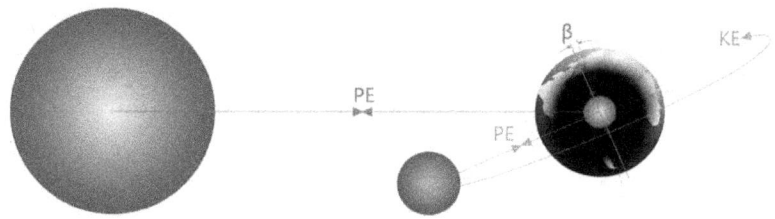

The angular tilt (β) between the two axes can be calculated thus:

β = sign(ω/ω_m) . ½.$\sqrt{[\,|Asin(\omega/\omega_m)|\,]}$

= 0.109553685228394 radians (**6.27696379369167°**)

11 Core Pressure

Core pressure refers to the pressure within any *mass* comprising more than 6 atoms; even a lump of metal. It is due to the same non-polar magnetism in the Quanta of which it is composed.

It is the same magnetic force that is attracting satellites to their force-centres and can be established using the force law generated first by Gilbert and again later by Newton:

$$F = G.m_1.m_2/R^2$$

11.1 Active Bodies

Active bodies such as stars, and planets with substantial lunar *mass*-populations:

1) *Gravity* and matter mobility will ensure that layer densities will always be greatest at the core and least at the outer surface (except for a crust).

2) There is insufficient potential pressure to generate fusion in stars and planets, so matter density greater than iron may be ignored in this calculation.

3) The inability for active planets and stars to generate fusion will ensure that the core density will be constant ($\Delta_i \approx 1$).

4) *Gravity* will ensure that density variation will be approximately linear throughout the mantle.

5) The core is always going to be more than 90% pure iron (≈ 8000 kg/m³), even in stars.

6) The radial modifier of planet crusts; i.e. those that have not melted, will be equal to one ($\Delta = 1$).

7) Atmospheric pressure of gas planets and stars should be added to all layers, but should not be included as a layer in the calculation because it will rotate at a different angular velocity to the body itself.

8) The more layers in the calculation, the more accurate will be the result. It is expected that a calculation result from seven or eight layers will provide an acceptable level of accuracy.

11.2   Inactive Bodies

The polar moment of inertia of galactic force-centres is totally unknown. They will be mostly iron, like all large celestial bodies, but their matter will never have been active during the current universal period. The radial modifier for the entire body can be assumed equal to 1; i.e. it is probably a pointless exercise performing this calculation for galactic force-centres.

For planets of significant *mass* with no moons of their own and orbiting close to their force-centre, the rules for active planets may be applied.

The results for a moon will be more speculative than with active bodies because there is no way of knowing whether they were once active somewhere else in the galaxy. However, if its surface matter density, total *mass* and its polar moment of inertia (through spin theory) are all known, a layer model of the body may be estimated using its radial modifier.

1) *Gravity* will ensure that layer densities will generally be greatest at the core and least at the outer surface.

2) There is insufficient potential pressure to generate fusion, so matter density greater than iron may be ignored in this calculation.

3) *Gravity* will ensure that density variation should be approximately linear throughout the mantle.

4) If the outer surface density is known, this may be used along with a body's average density to estimate a linear density variation from core to surface.

11.3 The Earth

An example calculation that yields a surprising, but understandable result has been performed for the earth.

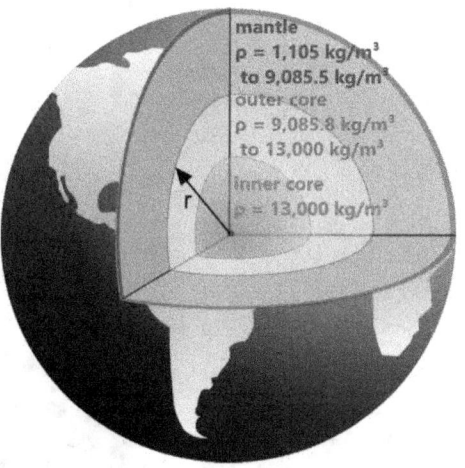

Most publicly available sources claim that the earth has a core density of somewhere around 13,000 kg/m³, which, given the incompressibility of iron, means it must contain a significant quantity of heavy elements.

Given the relatively low *mass* of the earth's core when compared with that of its mantle, reducing the core density to a more realistic value - closer to that of iron - changes the upper mantle density by only a small percentage (<1%). Therefore, the core density shall be assumed as stated above for this example calculation.

What do we know about the inside of the earth?

1) Its core is made of heavy elements, ρ_{core} = 13,000 kg/m³

2) Its surface comprises mostly water, $\rho_{surface}$ = 1250 kg/m³

3) Its inner core radius is ≈1215000 m

4) Its outer core radius is ≈3470000 m

5) Its upper mantle radius is ≈6363000 m

Note: 6km crust and 2km water depth

6) Its outer radius is 6371000 m

7) Its total *mass* is 5.95786E+24 kg

8) Its polar moment of inertia is 1.08209548E+37 kg.m²

CalQlata has created a calculator for this calculation that uses linearly variable layer densities for greater accuracy (see below). Using the above information and this calculator, the above results were achieved.

Whilst this discovery may be unexpected, it is perfectly logical.

It is currently claimed that mountain roots beneath the earth's crust eventually fall into the upper mantle, causing the crust to rise locally and

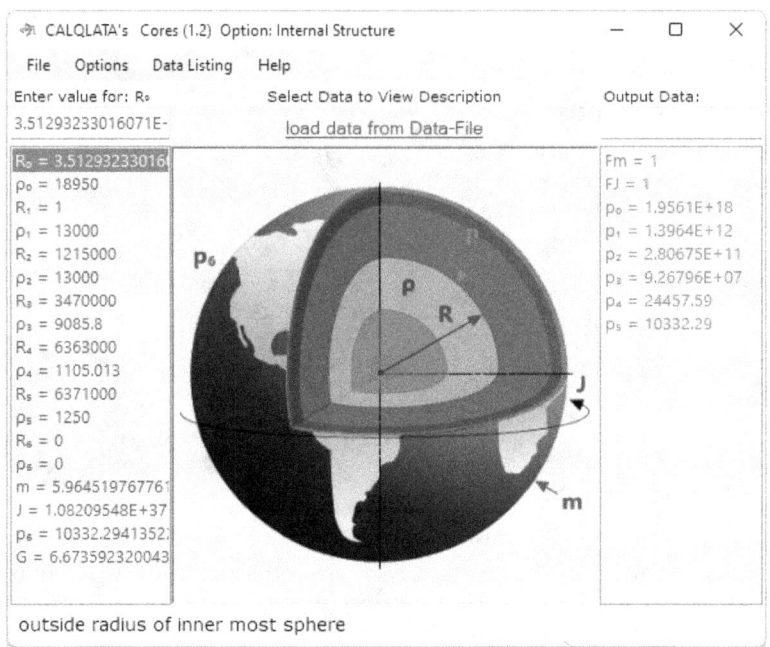

temporarily. This event would be difficult to explain if the density of the mantle were greater than, or even close to, that of the crust. It is not so difficult to understand, however, if the upper mantle is a hot, gaseous, pressurized, cauldron of matter with a lower density (ρ_4) than the crust it is supporting. Moreover, it is also easier to see where all the volcanic activity comes from. The same applies to subduction zones where crust material is pushed into the upper mantle. The sinking of the earth's denser, cooler continental crust into its mantle is actually aided by the relatively low density in the upper mantle and also generates the mantle plumes as it sinks towards the earth's core where it heats up and rises to the surface.

> R_0 to R_6 = outside radii for each spherical layer
> ρ_0 to ρ_6 = density at each spherical radius
> m = mass of the planet
> J = polar moment of area of the planet
> p_6 = atmospheric pressure at the planet's outer surface
> G = Newton's gravitational constant
> Fm = mass-factor (must equal 1)
> FJ = polar moment of inertia-factor (must equal 1)
> p_0 to p_5 = pressure at each spherical radius (R_0 to R_6)

The Mathematical Laws of Natural Science

This calculation procedure also tells us that the internal pressure is insufficient to support the overlaying crust. However, by applying 'PV=RT', we can establish the heat energy at the upper-mantle crust interface.

Given the atomic incompressibility of core matter (chapter 11.4.4), the core density must be close to that of iron (7870 kg/m^3) and the suggested upper mantle density (close to that of water) must be correct.

The Mathematical Laws of Natural Science

11.4 Calculations

Core-pressure can be defined using a modified version of Newton's formula:

$$p = G.m_1.m_2 / r^2.A \qquad \{m^3 / kg.s^2.m^4 = N/m^2\}$$

Where 'r' is the radius at which the pressure is to be calculated, m_1 is the *mass* inside 'r', m_2 is the *mass* outside 'r' and 'A' is the spherical surface area at radius 'r'.

Unlike an homogeneous body, an active celestial requires an iterative procedure to determine internal pressures due to their varying densities. But the same formulas apply.

11.4.1 Constant Density

A constant density calculation means that the entire body comprises matter of exactly the same density. The pressure at radius 'r' may be calculated thus:

The volume of the body:
$$V = {}^4/_3.\pi.r_o{}^3 \qquad \{m^3\}$$

The *mass* of the body:
$$m = V.\rho \qquad \{kg\}$$

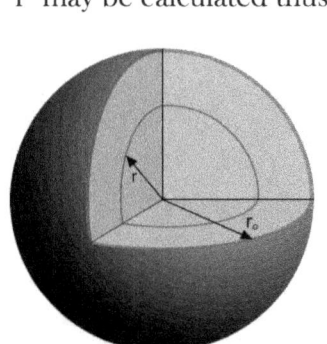

The volume of the *mass* inside 'r':
$$V_i = {}^4/_3.\pi.r^3 \qquad \{m^3\}$$

The *mass* of volume 'V_i':
$$m_i = V_i.\rho \qquad \{kg\}$$

The volume of the *mass* outside r:
$$V_o = V - V_i \qquad \{kg\}$$

The *mass* of volume 'V_o':
$$m_o = V_o.\rho \qquad \{kg\}$$

The spherical surface area at 'r':
$$A = 4.\pi.r^2 \qquad \{m^2\}$$

The internal pressure at 'r':
$$p = G.m_i.m_o / r^2.A \qquad \{N/m^2\}$$

And its polar moment of inertia (refer to chapter 10.3):
$$J = {}^2/_5.m.r^2 \qquad \{kg.m^2\}$$

11.4.2 Variable Density

A variable density calculation means that each layer comprises matter of different density. Before embarking on this calculation procedure, it is first necessary to determine the body's *mass* (m) and its polar moment of inertia using spin theory.

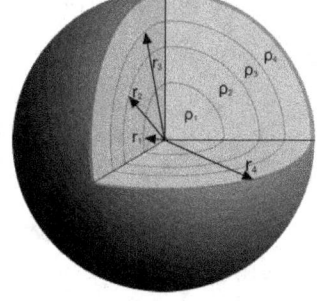

The pressure at radius 'r' may be calculated as follows:

The volume of each layer:

$$V_n = {}^4/_3 . \pi . (r_n^3 - r_{n-1}^3) \qquad \{m^3\}$$

The *mass* of each layer:

$$m_n = V_n . \rho_n \qquad \{kg\}$$

The polar moment of inertia of each layer (refer to chapter 10.3):

$$J_n = {}^2/_5 . m_n . (r_n^2 - r_{n-1}^2) \qquad \{kg.m^2\}$$

The following solution is for the pressure at 'r_2':

The *mass* inside 'r_2':

$$m_i = m_1 + m_2 \qquad \{kg\}$$

The *mass* outside 'r_2':

$$m_o = m_3 + m_4 \qquad \{kg\}$$

The spherical surface area at 'r_2':

$$A = 4.\pi.r_2^2 \qquad \{m^2\}$$

The internal pressure at 'r':

$$p = G.m_i.m_o / r^2.A \qquad \{N/m^2\}$$

And the polar moment of inertia of the total *mass*: $J = {}_1\Sigma^4 J_n$

Iterate radii (r_n) and densities (ρ_n) until the total *mass* ($m_i + m_o$) and polar moment of inertia (J) match the values defined using spin theory.

The Mathematical Laws of Natural Science

11.4.3 The Structure of Celestial Bodies

The calculation procedure described above may be used to estimate the structure inside a celestial body.

You simply build up a body of layers and iterate through radii and densities until you end up with calculated values for *mass* and polar moment of inertia that match the actual values found using spin theory.

There are a few pointers (rules) in performing this calculation that will help in setting up your model dependent upon whether or not the body is active; i.e. it generates internal heat.

11.4.4 Celestial Fusion

A typical fusion process - two beryllium atoms into one oxygen atom - is described in these images.

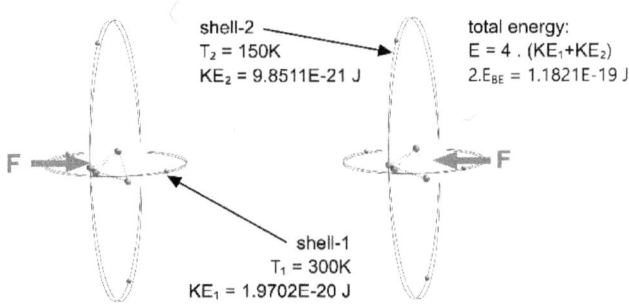

The gravitational (magnetic) force required to fuse these two atoms (F_m) must be greater than the electrical repulsion force (F_e) between the nucleic protons;

$F_e = k.e'^2/d^2$ where 'd' is the distance between the atoms at the time of fusion,

where, for example; d in beryllium at 300K is 2.0188337345E-10 m

resulting in a repulsion force (per proton) between nuclei;

$F_e = 0.019084371797$ N

The likelihood of fusion in the core of celestial bodies such as our earth and our sun, may be approximated by calculating the relative atomic forces thus:

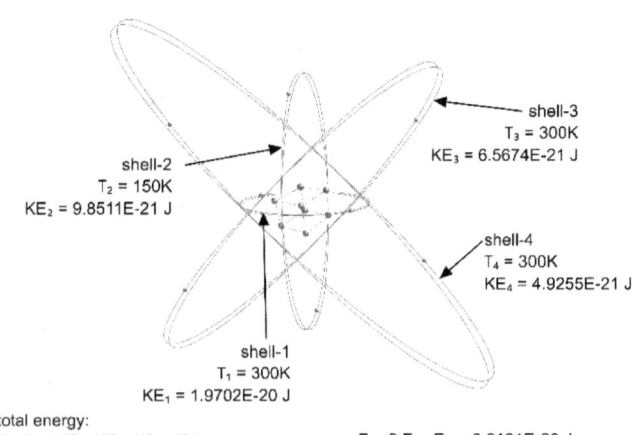

The Mathematical Laws of Natural Science

T = core temperature
d = distance between viscous iron atoms
e' = proton electrical charge at 'T'
F_e = repulsion force between protons at 'T'
A_p = cross-section area of a proton
m = body mass
r = calculation radius
ρ = average body mass density

$$F_m = G.m_1.m_2/R^2$$

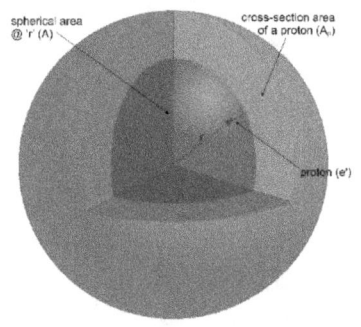

Sym.	earth	sun	units
T	6200	623316124.7	K
d	2.2817E-10	2.2817E-10	m
e'	2.9262E-21	2.94184E-16	C
F_e	1.478185E-12	1.494040E-02	N
A_p	9.91062E-30	9.91062E-30	m²
m	5.9579E+24	1.9885E+30	kg
R	6367444.5	447425129.2	m
ρ	5509.426596	5300	kg/m³

Inter-atomic magnetic force on a proton (F_p) inside a body of matter may be calculated thus;

$$F_p = F_m . A_p/A$$

fusion is only possible if $F_p > F_e$

The calculation results for the core of our earth are as follows:

r	2.2817E-10	0.01	1	4.89561425	10	100	m
A_r	6.54224E-19	0.125663706	12.5663706	301.178693	1256.63711	125663.71	m²
m_1	9.34618E-26	23.07783216	23077.8322	2.7078E+06	23077832	23078E+10	kg
m_2	5.95786E+24	5.95786E+24	5.9579E+24	5.9579E+24	5.9579E+24	5.9579E+24	kg
g	1.19806E-16	1.54012E-07	1.5401E-06	7.5398E-06	1.5401E-05	1.5401E-04	m/s²
F_m	7.13785E+08	9.17583E+17	9.1758E+18	4.4921E+19	9.1758E+19	9.176E+20	N
F_p	**1.08129E-02**	7.23662E-11	**7.2366E-12**	1.4782E-12	7.2366E-13	7.2366E-14	N

The calculation results for the core of our sun are as follows:

r	2.2817E-10	0.00015552	1	5	10	100	m
A_r	6.5422E-19	3.03925E-07	12.566371	314.159265	1256.6371	125663.71	m²
m_1	9.3462E-26	8.35025E-08	22200.5881	2775073	2.22006E+07	2.22006E+10	kg
m_2	1.9885E+30	1.9885E+30	1.9885E+30	1.9885E+30	1.9885E+30	1.9885E+30	kg
g	1.1981E-16	2.30411E-10	1.4816E-06	7.4079E-06	1.4816E-05	1.4816E-04	m/s²
F_m	2.3823E+14	4.58172E+20	2.9461E+24	1.4731E+25	2.9461E+25	2.9461E+26	N
F_p	**3.6089E+03**	1.49404E-02	2.3235E-06	4.647E-07	2.3235E-07	2.3235E-08	N

which reveals that **fusion** may have occurred in the atoms at their current core temperature within a 5-metre radius of the earth's core and in the atoms within a 0.0002-metre radius of the sun's core in the distant past. But given that fusion is not a continual process, the energy released at that time has long ago been radiated by the celestial bodies concerned. In reality, as radius 'r' increases, F_e reduces with decreasing temperature, but so too does F_m with reducing matter density, altering the result, although only slightly.

The Mathematical Laws of Natural Science

12 The Universe

The universe is an empty space larger than the outermost reach our galaxies will achieve during this universal period; 4.42755E+23m.

Its mass is such that it will generate the core pressure in a single spherical body sufficient to cause two neutrons to make contact, thereby initiating a nuclear chain reaction, that will blast more than 99% of its matter into outer space. Whilst most of this ejected matter will then travel outwards with an initial velocity (u = 1.7735E+06 m/s), after this explosion ('Big-Bang') residual matter will re-accrete into a single body that we will call the great attractor.

The outward travel of all universal matter will be slowed due to the potential energy (gravitational acceleration) induced by the great attractor.

Any EME emitted during the 'Big-Bang', which will have travelled at light-speed (c), will be more than 1.3E+26m from the great attractor by now. Whereas the galaxies were initially ejected at 0.59% of 'c' and has today slowed to 0.077% of 'c' due to the magnetic attraction of the great attractor. Therefore, the heat energy detected in outer-space during the Cobe project couldn't possibly be the heat left over from the last 'Big-Bang'; it can only be that radiated by all of the active celestial bodies in the universe.

The universe is one big energy generator. This energy is generated by all universal orbits (celestial and atomic), which is converted into heat energy through spin-friction in *bright* celestial bodies, where it is stored in neutrons. These neutrons will be the source of the energy released during the next 'Big-Bang'.

This process can repeat itself, unaided, indefinitely.

All the celestial bodies in the universe were ejected at the same time with similar energies per unit mass, which means that all celestial bodies are a similar distance from the 'Big-Bang' (and the great attractor); they are at the periphery of an ellipsoid that is today about 4.3531E+23m from its centre (the great attractor); and will stop expanding at a little more than 4.42755E+23m in just under 16 billion years from initiation, after which it will return over the same period to accrete into another ultimate body.

The mass of our galaxy is approximately 1.76572E+41 kg

All the matter in the universe is greater than 4.6868788E+48 kg

The Mathematical Laws of Natural Science

Therefore, if we are orbiting within an average galaxy, there are more than 26.54 million galaxies in the universe, all of which are at the periphery of an ellipsoid approximately 8.71E+23 metres across today. That we are at the periphery of an ellipsoid is borne out by Hubble's discovery of red-shift in almost all of the galaxies, and that the further they are from us the faster they are travelling away from us. And if there are approximately 200 million stars in all the galaxies (chapter 14), there are around 5.3E+15 stars in the universe; but not all of them will be *bright*!

12.1 The Age of the Universe

The only two unknowns for an *actual* solution to all the properties of our universe are; the final expansion (R_o) and the great attractor mass (m_1), either of which may be found thus:

$G \cdot \sqrt{[m_2 / 2.E.u^2]}$ {m/kg}
$= R_o / m_1$ (if you know one (R_o or m_1), you can find the other)

Isaac Newton's constants of proportionality and motion for orbits can also be applied to universal linear expansion like this:

$K = (2\pi)^2 / G.m_1 = (2\pi)^2 / 2.R_o.(u/2)^2 = 2.\pi^2 \cdot t_o^2 / R_o^3$ {s²/m³}

$h = R \cdot v/2$ (a variable) {m²/s}

$E.(K.h)^2/m_2 = 8.\pi^4$ (a variable)

$2.E/m_2 = (G.m_1 / h)^2$ {m²/s²}

$G \cdot \sqrt{[m_2 / 2.E.u^2]} = R_o / m_1$ {m/kg}

$m_2.u^2 / E = -2$

universal mass: $4.68687882273807E+48 < m^U \leq 1E+49$ kg

energy released ('Big-Bang'): $E = 7.35440361613308E+60$ J

initial velocity: $u = \sqrt{[2.E/m^U]} = 1.7734984104391E+06$ m/s

Assuming **NASA's *estimates*[#]**, we can calculate a universal period based upon the known mass of Hades ($m_h = 1.76572E+41$ kg) as follows ...

current age: $t = 4.34548152E+17$ s (***13.77 bn-yr***)[#]

current velocity: $v = $ ***230000 m/s***[#]

great attractor mass: $m_1 = 1.04336261407224E+46$ kg^{##}

dynamic acceleration: $a = -u/t = -3.55196174079941E-12$ m/s²

gravitational acceleration today: $g = -G.m_h/R^2 = -3.6745242336E-12$ m/s²

distance travelled today: $R = (v^2-u^2) / 2.a = 4.35308265895621E+23$ m

ultimate distance travelled: $R_o = u^2 / 2.a = 4.42754855113119E+23$ m

time at R_o: $t_o = u/a = 4.99301101717368E+17$ s (15.8219 bn-yr)

universal period: $t = 2.t_o = 31.64379$ billion years

The Mathematical Laws of Natural Science

\# *The current age of the universe according to NASA (13.77 billion years) is simply guesswork based upon a myth (Relativity), so it cannot be taken seriously. NASA's estimated velocity of our galaxy (230000 m/s) should be simpler to identify as it is based upon red-shift values, however, not too long-ago NASA proposed 600000 m/s, so this value must also be considered highly suspect.*

\#\# *This value is selected for the calculation simply to reflect NASA's estimates for the Milky Way's current velocity and age. If, however, accurate values for either R_o or m_1 can be found, we would be able to define <u>everything</u> about our universe without NASA's estimates.*

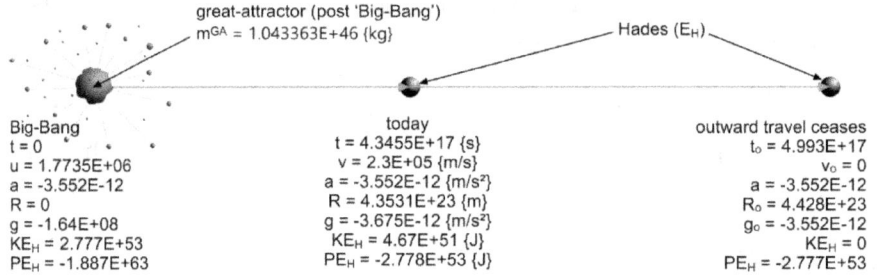

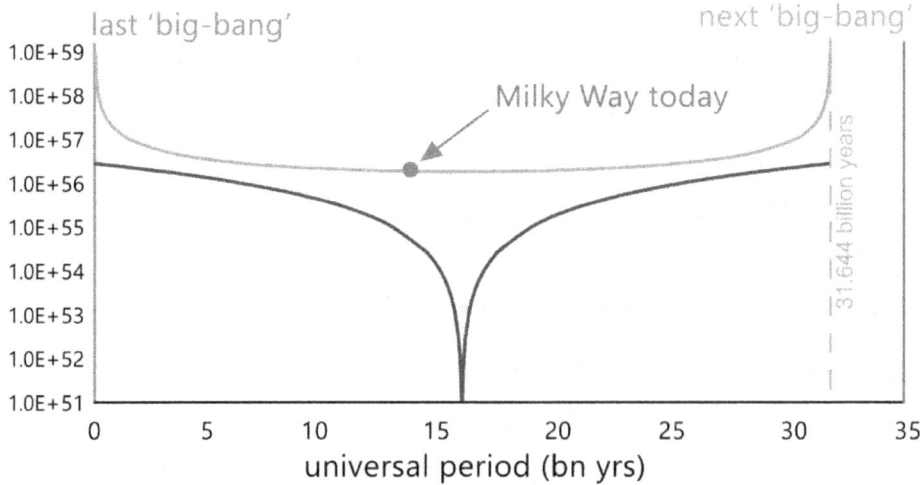

12.2 The Big-Bang

The 'Big-Bang' was the release of neutron energy from a single body-mass, sitting alone in empty space (the ultimate body), more than 99% of which was ejected at a little less than 0.6% of light-speed.

The ultimate body was the final magnetic (gravitational) accumulation of all the matter in the universe. It was greater than 4.7E+48 kg, which is the minimum perfectly spherical mass required to overcome the coupling ratio. The neutrons in all this matter were the source of the energy released at its final destruction; the '*Big-Bang*'; $\gtrsim$ 7.3544E+60 Joules.

During the 'Big-Bang', the entire mass was decimated, apart from a residue of 1.040053347E+46 kg of rubble that re-accreted to form the great attractor.

This release of energy was exactly the same as that which occurs in an atom bomb, or a nuclear reactor, both of which release neutron energy from radioactive matter in a chain reaction. The release of energy in the 'Big-Bang', however, was from ordinary rock!

12.3 The Universal Process

Step-1: galactic outward travel (from the last 'Big-Bang') stops due to the magnetism of the great attractor, following which all galaxies gradually return to the great attractor, increasing its mass until it achieves the mass of the ultimate body (critical mass).

Step-2: on achieving critical mass (> 4.68688E+48 kg), core pressure causes neutrons to make direct contact, initiating a nuclear chain reaction splitting ≈3% of the neutrons into their component parts, releasing their stored energy in the process.

Step-3: most of the matter of the ultimate body will be ejected into empty space at a velocity of ≈1.7735E+06 m/s, the largest of which becoming galactic force-centres, collecting nearby smaller bodies either into orbit (galactic satellites) or into themselves augmenting their mass.

Step-4: the residual mass (left over from the 'Big-Bang') re-accretes into a single mass of ≈0.2% of the ultimate body; the great attractor.

Step-5: orbital precession (orbital axis rotation) causes many galactic satellites (stars) to impact, creating galactic comets and galactic asteroid belts.

Step-6: stars gradually collect these comets as sub-satellites (planets), generating internal frictional heat in stellar force-centres, which gradually rises with increasing planetary mass

Step-7: Eventually, many stars achieve the neutronic temperature (T_n) in their cores, at which time, fissionable energy is released; they become bright (bright stars).

Step-8: planets simultaneously trap galactic and solar comets as sub-sub-satellites (moons), that generate internal frictional heat in their planetary force-centres. Some planets eventually achieve the neutronic temperature (T_n), at which time, fissionable energy is released; they become bright (bright-planets).

Step-9: return to Step-1

The Mathematical Laws of Natural Science

12.4 Calculations

The mass of the ultimate body is such that the core pressure is sufficient to overcome the electrical repulsion between two adjacent protons. When this is achieved, two neutrons will make contact, causing them to split; the start of a nuclear chain reaction. In a body this massive, the internal chain reaction will take some time to reach the body surface. Destruction will take place before this occurs and the body parts will be flung into outer space, releasing about 3% of its neutron energy; 'Little Boy'. It works like this:

Body mass:
$$m_u = k.e^2 \div (G.m_n.\varphi) + 1 = 4.68687882273808E+48 \text{ kg}$$

Number of proton-electron pairs:
$$N^{\underline{o}} = m_u/m_n = 2.80059013353655E+75$$

Universal mass:
$$m_u = \text{ultimate body mass - great attractor mass}$$
$$= 4.6868788E+48 - 1.0400534E+46$$
$$= 4.676478E+48 \text{ kg}$$

Neutron energy:
$$E_n = 1.63785606465701E-13 \text{ Joules (refer to chapter 5.3.2)}$$

Number of neutrons:
$$N^{\underline{o}}_n = 2.80059013353655E+75 \, . \, (1-1/(1+\psi))$$
$$= 1.49675415620709E+75$$
where $\psi = 1.147962$ (e.g. iron)

Total energy store inside ejected matter:
$$E_t = N^{\underline{o}}_n.E_n$$
$$= 2.45146787E+62 \text{ Joules}$$

Energy released in ejected matter (based upon Little-Boy; 3%):
$$KE = 2.45146787E+62 \times 3\%$$
$$= 7.35440362E+60 \text{ Joules}$$

Initial velocity of ejected matter:
$$V_0 = \sqrt{[2.KE / m]}$$
$$= 1,773,498..4104 \text{ m/s}$$
$$= 0.59\% \text{ of '}light\text{-}speed\text{'}$$

The Mathematical Laws of Natural Science

13 Celestial Bodies

'Celestial body' is a term commonly used to describe all matter that exists in space. It refers to; the Great Attractor, galactic force-centres, stars, planets, moons, comets, asteroids, meteorites, etc.

All celestial bodies originated from the ultimate body immediately after the last 'Big-Bang', and were created during previous universal periods through fusion, fission and planetary core chemistry.

Stars, planets and some moons only appear as they do today because of the heat energy generated in them through spin-friction.

The core of the universe's largest celestial bodies – the ultimate body, the great attractor and the galactic force-centres – which are cold because they are not in elliptical orbits, are where all atomic fusion is generated.

The universe's hottest bodies; bright stars and planets, are the only places where neutrons are created naturally from fissionable energy, the heat for which is generated by planetary spin.

Gas planets are satellites that have collected sufficient sub-satellite mass to generate the heat needed to melt their crusts via planetary spin. They tend to be furthest from their force-centre's gravitational influence.

Life-giving planets are satellites that have collected sufficient sub-satellite mass to generate internal heat from planetary spin to create continental drift in their solid crusts. They tend to orbit between the barren and gas planets.

Barren bodies are those that generate little or no internal heat through planetary spin. They tend to be closest to their force-centre, making it difficult for them to trap galactic comets due to their force-centre's gravitational influence.

All matter, from individual atoms upwards, is bound to celestial bodies by gravity. Universal space outside celestial body atmospheres contains nothing, other than electro-magnetic energy; there is no such thing as dark matter.

13.1 The Ultimate Body

The ultimate body is the accretion of all universal mass at the end of a universal period.

The ultimate body comprises all the matter in the universe. It is the source of a 'Big-Bang' due to neutron-neutron interaction at its core from core pressure.

It is the re-accretion of all universal matter, having been collected by the gravitational attraction of the great attractor. It will sit alone in empty space. Its mass will be the minimum required by a perfect sphere to overcome the coupling ratio (φ).

Because it comprises all of the matter in the universe, it will have a similar average density, albeit increased due to core pressure. Therefore, we can estimate the average density of the ultimate body to be approximately that of lead; the heaviest of the stable elements.

Its properties may be estimated as follows:

    average density: $\rho \geq 11000$ kg/m^3 (estimated)

    mass: $m > 4.68688\text{E}+48$ kg

    surface radius: $r = 4.66803543\text{E}+14$ m

    surface area: $A = 2.738282\text{E}+30$ m^2

    surface gravitational acceleration: $g = 1.435407168\text{E}+09$ m/s^2 (1.464E+08 g)

    proton-electron pairs: $N^{\underline{o}} > 1.30384\text{E}+75$

    neutrons: $N^{\underline{o}} > 1.49675\text{E}+75$

    stored energy: $E = 2.4514679\text{E}+62$ J

Given that an ultimate body comprises all the universal stars (<1% of universal mass @ ≈6000K) and all the universal galactic force centres (>99% of universal mass @ ≈40K), it will have a temperature of approximately 100K.

Due to its mass and low temperature, the ultimate body is a major source of universal fusion in much of its matter.

13.2 The Great Attractor

The great attractor is slowing down universal expansion, which will eventually cease and re-accrete all universal matter as an ultimate body.

The great attractor is the residual (rubble) left over from the last 'Big-Bang', which will have quickly re-accreted into a single mass. It sits alone in empty space at the centre of the universe. Its mass is sufficient to slow down the outward travel of all ejected matter to their current velocity ($\approx$230000 m/s).

Because it comprises the same matter as the rest of the universe, it will consist of similar matter; predominantly iron, but its density will be greater due to core pressure, say that of the heaviest stable element; lead.

Its properties may be estimated as follows:

   average density: $\rho \approx 11000$ kg/m^3 (estimated)

   mass: m = 1.04336261407224E+46 kg (estimated)

   surface radius: r = 6.51680487174181E+13 m

   surface area: A = 5.33677998450059E+28 m^2

   surface gravitational acceleration: g = -1.639553184E+08 m/s^2 (16,718,821 g)

   proton-electron pairs: N° > 2.89330922153997E+72

   neutrons: N° > 3.32140904057746E+72

   stored energy: E = 5.43998994031642E+59 J

   constant of proportionality: K = 5.66976153229753E-35 s^2/m^3

It is expected that the temperature of the matter comprising the great attractor at the start of a universal period would be approximately 100K, falling to about 40K at its termination.

Due to its mass and low temperature, the great attractor is a significant source of universal fusion in much of its mass. It is also the reason we cannot see it in the night sky (dark).

13.3   Galactic Force-Centres

Galactic force-centres are cold (dark) because they are not in elliptical orbits.

Galactic force-centres are the largest pieces of matter ejected from the ultimate body during the last 'Big-Bang'. And because they all originate from the same source as the great attractor, they comprise similar matter, but they are much smaller; 0.001697721% so their densities will be only a little more than iron (≈9000 kg/m^3).

They are dark because they are cold (≲70K), and they are cold because they are not in elliptical orbits; they cannot generate internal frictional heat. All galactic force-centres are travelling away from the great attractor in a straight-line. Their initial velocity, immediately after the last 'Big-Bang', was ≈1.772E+06 m/s and is today about 230000 m/s; they are slowing down.

All galaxies are travelling away from the great attractor at a similar velocity, and are a similar distance, because they were all ejected at the same time under the same release of energy. The great attractor's gravitational attraction will bring them to a halt at ≈4.423E+23 metres in 15.825 billion-years, after which they will all travel back to the great attractor where they will re-accrete into another ultimate body.

Due to their mass and low temperature, galactic force-centres are a source of universal fusion in their cores. And because they are cold (dark) they radiate very low EME. Which is why we cannot see them; *they are not black holes.*

13.4 Stars

Fission (and spin) is the heat (energy) source of a bright star.

Stars are galactic satellites. Like every other celestial body, stars were originally ejected from the ultimate body during the last 'Big-Bang'. They therefore comprise the same matter as all other celestial bodies. They were originally attracted to the nearest galactic force-centre, some of which will have been consumed by the force-centre, augmenting its mass, but others, with apposite kinetic and potential energies, will have become orbital.

All galactic satellites are stars, but not all of them are *bright*.

A *bright* star is a galactic satellite that has collected sufficient sub-satellite mass to create the internal frictional heat - through planetary spin - necessary to generate fission in its core atoms.

Brightness can occur in planets if they are sufficiently large and collect the sub-satellite mass needed to generate fission. These are not stars; they are *bright* planets. There are no such things as binary stars; only one can be a star (a force-centre), the other must be a planet (a satellite).

All stars were initially cold. Over time, precession has created numerous galactic comets that were collected by the stars as sub-satellites (planets) gradually heating them through planetary spin. When sufficient core heat is being generated, fission will occur in their core atoms.

The process working within all *bright* stars and planets is neutron creation, fission and EME radiation. Their core temperature is neutronic ($\underline{T}_n$); the highest possible in nature. The surface temperature of bright stars will vary with size. Larger bodies will be colder and radiate lower EME.

Because of the fission being generated within a bright body, its core atoms are gradually breaking down to proton-electron pairs, and because this occurs in pairs (two electrons per shell), the atoms at the surface of all bright bodies are hydrogen and helium.

Whilst their body mass does not alter with time, their viscous matter will decline as their gaseous matter increases; they increase in size with age. Eventually, there will be insufficient viscous body mass to maintain neutronic temperature at their cores; core fission will cease. During this process, their surface temperature will increase, but heat intensity will reduce.

13.5 Planets

Planets are stellar satellites. They originate from stellar collisions, the rubble from which become galactic comets that are trapped into solar orbits as they pass through the galaxy. They therefore comprise the same matter as all other celestial bodies, however, they lack the compression and uniformity of much larger bodies and will therefore have lower, more variable densities.

Planets provide their stellar force-centres with internal frictional heat through planetary spin. Once sufficient planetary mass has been collected, the stellar force-centre will achieve the neutronic temperature at its core, making it bright.

The planets orbiting non-bright stars will be cold for a number of reasons;
1) They will receive negligible radiated EME from their star
2) Their star has insufficient mass to collect substantial planetary mass
3) If the star is small, its ability to trap planets will be low
4) If a small star traps a relatively large planet, the star will be pulled into a localised orbit, preventing the generation of internal frictional heat in the star's core.

The orbital organisation about a bright star is as follows:

13.5.1 Inactive

Inactive planets are those that cannot trap sub-satellites of their own due to the gravitational influence of their force-centre's mass. They tend to be those orbiting close to their star (e.g. Venus and Mercury). Whilst they may have hot surfaces due to their proximity to their stellar force-centres radiated EME, they will generate little or no internal heat.

13.5.2 Active

Active planets are those that attract sufficient sub-satellite mass to generate internal heat from spin-friction and thereby create continental drift in a cooler outer crust.

These planets tend to orbit between the inner barren planets and the outer gas planets. It is where life will be found.

13.5.3 Gaseous

Gas-planets are those that have trapped sufficient sub-satellite mass to generate the internal frictional heat required to melt their crusts. These planets tend to orbit furthest from their stellar force-centre's gravitational influence and are therefore too distant from their stellar force-centre to receive significant radiated heat. Almost all of their heat energy comes from spin-friction.

The atmosphere surrounding gas planets will be from molten rock, minerals and elements in their structure.

13.6   Moons

Moons are planetary satellites. They originate from stellar and planetary collisions, the rubble from which becomes galactic or solar comets that are trapped into planetary orbits as they pass through the galaxy or solar systems.

Moons provide their planets with internal heat energy through spin. But because they are unlikely to attract satellites of their own, they will normally be cold. However, the gravitational influence on moons from massive planets can create substantial internal disruption (e.g. Io).

It cannot be overstated, however, that our own moon is responsible for generating 70% of our own planet's surface heat through spin friction. Without it, earth's surface temperature would fall to 90K (-183 °C), and continental drift would cease (refer to chapter 10.2).

13.7   Comets & Meteorites

Comets and meteorites are simply orbiting bodies that have yet to be trapped or consumed by their force-centres or planets. They are usually rubble ejected from planetary collisions and therefore comprise the same matter; rock and water, as all other celestial bodies.

The Mathematical Laws of Natural Science

14 The Milky Way

The mass of our sun (1.9885E+30 kg) is approximately 745.5 times greater than the sum of all its planetary mass (2.6676E+27 kg). Therefore, our stellar force-centre constitutes 99.866% of the mass of our solar system

If all the other solar systems orbiting Hades are similar to ours, and the stellar:galaxy mass ratio is similar to our solar system there are about 1.19E+08 - almost 200 million - solar systems in our milky way.

Our solar system is in the outer reaches of the milky way, approximately 2.54E+20 metres from Hades.

To summarise, the milky way:

> Has a total mass of ≈1.769697E+41 kg
>
> Its galactic force-centre has a mass of 1.76572E+41 kg
>
> Comprises approximately 200 million stars (most being dark).
>
> Perigee distance of 2.4653729E+20 m
>
> Apogee distance of 2.54525098196E+20 m
>
> Has an orbital eccentricity of 0.0159417437512
>
> Is currently 4.35308E+23 metres from the great attractor
>
> Is currently travelling away from the great attractor at ≈230000 m/s

The Mathematical Laws of Natural Science

14.1 Hades

Based upon NASA's estimates of our sun's galactic orbit, we can estimate the properties of our own galactic force-centre, which I have named 'Hades' to save me from repeatedly referring to '*milky-way galactic force-centre*'.

- average density: $\rho \approx 9000$ kg/m³ (estimate)
- mass: m = 1.76572018982E+41 kg
- distance from great attractor: 4.35308E+23 m
- linear velocity: 230000 m/s (est.)
- surface radius: r = 1.673132E+12 m
- surface area: A = 3.517794E+25 m²
- surface gravitational acceleration: g = 4.209408E+06 m/s²   (4.3E+05 g)
- constant of proportionality = 5.68780171912379E-35 s²/m³
- polar moment of inertia: 1.97716E+65 kg.m²
- potential energy: PE = -2.824359E+53 J
- kinetic energy: KE = 4.67033E+51 J
- spin energy: SE = -8.75632E+48 J
- spin-rate: ω = -9.41141E-09 c/s
- temperature $\lesssim$ 70K

Hades is in a linear orbit (e = 1), therefore, it cannot generate internal heat energy (through spin friction); it is cold. It only holds and dissipates the heat it held prior to the 'Big-Bang', its radiated EME will be low; it is dark:

EME radiation: $\lambda \approx$ 4.704644E-04 m
$A \approx$ 2.509237E-08 m
$f \approx$ 6.372266E+11 /s
$E \approx$ 4.597168E-21 J

And given the sun's orbital radius ($\approx$2.5E+20 m) and the size of an iron atom, searching for Hades in the night sky would be like looking for a black atom in the centre of an eleven-metre diameter black disk. That's why we can't see it; but it doesn't mean it isn't there! We know it exists because it is a fundamental law of nature that every orbital system *must* have a force-centre.

The Mathematical Laws of Natural Science

14.2 Our Sun

Our sun is a typical bright star. It has sufficient mass and collected sufficient sub-satellite mass to generate the internal frictional heat necessary to generate fission in its core atoms.

Like all stars, it was originally a cold, dark, lone galactic satellite comprising similar matter to all other celestial bodies; it had no sub-satellites of its own. But as time (< 13.77 billion years) passed, it collected its own satellites from galactic comets and began to heat up. As time continues to pass, it will collect more satellites (planets) increasing its internal frictional energy, but because the fissionable energy generated in its core is so much greater than the spin energy, its radiated heat will change little over the remaining universal period.

Our sun generates more than 28.12 times as much fissionable energy than it generates from planetary spin.

We know quite a lot about our star simply from its planetary orbits:

mass = 1.9885E+30 kg

density (viscous matter) ≈ 5500 kg/m³

viscous diameter = 8.8386E+08 m

orbital period = 7.258248E+15 s

orbital velocity = 213444.9459 to 220360.5621 m/s

orbital radius = 2.4653729E+20 to 2.54525098196E+20 m

constant of motion = 5.43270958E+25 m²/s

constant of proportionality = 3.3502574460E-30 s²/m³

spin energy = 1.6010041076E+35 J

polar moment of inertia: 3.90008E+46 kg.m²

spin-rate: ω = 2.86533E-06 °/s

core temperature = 623316124.717178 K

surface temperature = 5788 K

EME radiation ≈ 9.9252916097005E+33 J

fissionable energy generation ≈ 4.50156641838639E+36 J

solar power = 4.75535480329.58E+48 J/s (Watts)

The Mathematical Laws of Natural Science

It is currently claimed that our sun is creating elements from hydrogen (H⁺) through fusion and that it is [apparently] growing in size with age.

The problem with this scenario is that fusion *increases density* and therefore *reduces size*. Moreover, why is Hades cold if fusion generates heat, as Hades is far more likely to generate fusion than our sun.

Fissionable decay generates most of the sun's radiated heat energy and is therefore responsible for its hot hydrogen-helium atmosphere. The only reason we can see the atmosphere is because the hydrogen and helium at its surface comprises proton-electron pairs, which *are* capable of emitting electro-magnetic radiation. Our sun's atmosphere would be invisible (and cold) if its surface was lone protons (H⁺).

14.3 Planetary Temperature

We know that the sun radiates its heat into outer-space, achieving 90K at the earth's surface; refer to https://www.calqlata.com/Science/Atmosphere.html. Using this as our datum, we can establish the heat received from the sun for any other planet in the solar system thus: $T_r = 90 \cdot (R_E/R_P)^2$

	m:m_p (%)	R (m)	T_r (K)	T_s (K)
Mercury	N/A	6.9815E+10	427.15	700 #
Venus	N/A	1.0893E+11	175.44	<500
Earth	1.23%	1.5209E+11	90	300
Mars	0.00000189l%	2.4921E+11	33.52	227
Jupiter	0.0207%	8.1561E+11	3.13	>1500K ##
Saturn	0.0247%	1.5011E+12	0.92	>1500K ##
Uranus	0.0102%	2.9977E+12	0.23	>1500K ##
Neptune	0.0210%	4.5483E+12	0.10	>1500K ##
Pluto	11.88%	7.3711E+12	0.04	40

m:m_p is the percentage (ratio) of lunar mass to planetary mass
R is the planet's orbital apogee
T_r the sun's calculated contribution to the planet's average surface temperature
T_s the apparent temperature at the surface of the planet
Notes:
the hot side of the planet
the expected melting temperature of surface rock

The surface temperature of each planet comes from the heat radiated by the sun plus that generated within the planet itself from spin-friction. We can predict the expected surface temperature using the earth as the datum:

Earth's Core; 6000 K

Earth's Surface (radiation): 90 K

Earth's Surface (internal energy): 210 K

Planet's surface temperature from radiation: $T_r = 90 \cdot (R_E/R)^2$

Planet's core temperature from spin-friction: $T_c = 6000 \cdot (\delta\omega_E/\delta\omega)$

Planet's surface temperature from spin-friction: $T_s = 210 \cdot T_c/6000 \cdot \sqrt{[r/r_E]}$

Planet's surface temperature: $T = T_r + T_s$

where:
R_E = Earth's orbital apogee
R_P = the planet's orbital perigee
R = planet's orbital radius
$\delta\omega_E$ = Earth's core:mantle rotation
$\delta\omega$ = planet's core:mantle rotation
r_E = Earth's body radius
r = planet's body radius

14.4 Mercury

Sym	Value	Units	Sym	Value	Units
Orbital Dimensions			**Core-Mantle Dimensions**		
t	7600521.6	s	rc	458566.0805	m
R^P	4.600120E+10	m	r	2439700	m
R^A	6.981450E+10	m			
Orbital Shape			**Planetary Spin**		
a	5.790785019E+10	m	Δ	0.812862196	
e	0.02056137492		J	5.193084E+35	kg.m²
b	5.667054609E+10	m	ω_0	8.2667817E-07	c/s
p	5.545967920E+10	m	E_0	1.774469E+23	J
f	4.60012E+10	m	ω_1	1.490135E-06	c/s
x'	1.190665019E+10	m	E_1	5.765629E+23	J
A	1.030966877E+22	m²	ω_2	1.239801E-06	c/s
L	3.599700958E+11	m	E_2	3.991161E+23	J
K	2.974914364E-19	s²/m³	ω_3	0	c/s
v^P	5.897421240E+04	m/s	E_3	0	J
g^P	-6.271146496E-02	m/s²			
v^A	3.885846816E+04	m/s			
g^A	-2.722663689E-02	m/s²			
Masses					
m_1	1.9885E+30	kg	mc	2.597868E+21	kg
m_2	3.301100E+23	kg	m_m	3.275121E+23	kg
Satellite Performance			**Core-Mantle Spin**		
F^P	-2.070168170E+22	N	Δ_c	1	
PE^P	-9.523022001E+32	J	Δ_m	0.830714063	
KE^P	5.740543129E+32	J	J_c	2.185149E+32	kg.m²
F^A	-8.987785104E+21	N	J_m	5.190899E+35	kg.m²
PE^A	-6.274777265E+32	J	Ω_c	4.030034E-05	c/s
KE^A	2.492298393E+32	J	E_c	1.774469E+23	J
E	-3.782478872E+32	J	ω_m	-1.490449E-06	c/s
h	2.712884540E+15	m²/s	E_m	-5.765629E+23	J
			$\delta\omega$	3.880989E-05	c/s
			δE	-7.540098E+23	J
Core Pressure @ <1K					
pm	9.687E+22	N/m²	pe	7.646E+27	N/m²
Calculated Temperatures					
			T_r	600.62	K
			T_c	3356.22	K
			T_s	72.7117	K
			T	673.33	K

Mercury
Input Data; Estimate
^{P} = perigee; ^{A} = Apogee; $_c$ = core; $_m$ = mantle
$_0$ = orbital PE; $_1$ = force-centre spin KE; $_2$ = total; $_3$ = sub-satellite KE
Refer to chapter 14.3 for temperatures.

Mercury

ρ_{ave} = 5427.012135 kg/m³

Mercury's matter is similar to that when it was ejected during the last '*Big-Bang*'. Its density and 'Δ' value show it to be an iron-rich planet with an internal structure much more evenly distributed than is the case for Venus or the earth. It will have minimal surface activity as it generates little internal energy; 2.284 J/kg.

Mercury has little or no surface activity (volcanoes and earthquakes), due to the minimal internal activity, so its appearance will never change during its existence as a planet.

Because its internal frictional heat comes from E_0 & E_1 only (it has no satellites of its own) it has relatively low core-mantle activity, resulting in insignificant internal frictional heat or magnetic field. This is also the reason why the surface of the planet's '*far-side*' is cold, despite being so close to the sun.

Mercury receives 600K of its surface temperature from the sun, and generates 73K from spin-friction. Its highest surface temperature is said to be 700K.

There is little to distinguish Mercury from, say, Pluto other than its proximity to its force-centre, which is the reason Mercury has acquired no satellites of its own.

The Mathematical Laws of Natural Science

14.5 Venus

Sym	Value	Units	Sym	Value	Units
Orbital Dimensions			**Core-Mantle Dimensions**		
t	*19413907.2*	s	r_c	1098857.372	m
R^P	*1.074770E+11*	m	r	6051800	m
R^A	1.089346E+11	m			
Orbital Shape			**Planetary Spin**		
a	1.082057842E+11	m	Δ	0.68123191	
e	0.006735168186		J	3.309127E+37	kg.m²
b	1.082033299E+11	m	ω_0	3.2364352E-07	c/s
p	1.082008757E+11	m	E_0	1.733075E+24	J
f	1.07477E+11	m	ω_1	1.232980E-07	c/s
x'	7.287841549E+08	m	E_1	2.515333E+23	J
A	3.678247729E+22	m²	ω_2	-2.992369E-07	c/s
L	6.798692829E+11	m	E_2	-1.481541E+24	J
K	2.974914364E-19	s²/m³	ω_3	0	c/s
v^P	3.525676706E+04	m/s	E_3	0	J
g^P	-1.148825844E-02	m/s²			
v^A	3.478502382E+04	m/s			
g^A	-1.118288438E-02	m/s²			
Masses					
m_1	1.9885E+30	kg	m_c	3.630373E+22	kg
m_2	*4.867370E+24*	kg	m_m	4.831066E+24	kg
Satellite Performance			**Core-Mantle Spin**		
F^P	-5.591760450E+22	N	Δ_c	1	
PE^P	-6.009856379E+33	J	Δ_m	0.695161171	
KE^P	3.025166886E+33	J	J_c	1.753452E+34	kg.m²
F^A	-5.443123593E+22	N	J_m	3.307374E+37	kg.m²
PE^A	-5.929443189E+33	J	Ω_c	1.405972E-05	c/s
KE^A	2.944753696E+33	J	E_c	1.733075E+24	J
E	-2.984689493E+33	J	ω_m	-1.233307E-07	c/s
h	3.789291553E+15	m²/s	E_m	-2.515333E+23	J
			$\delta\omega$	1.393639E-05	c/s
			δE	-1.984608E+24	J
Core Pressure @ <1K					
pm	1.428E+24	N/m²	pe	3.638E+28	N/m²
Calculated Temperatures					
			T_r	172.02	K
			T_c	1205.1019	K
			T_s	41.1199	K
			T	213.14	K

Venus
Input Data; Estimate
^{P} = perigee; ^{A} = Apogee; $_c$ = core; $_m$ = mantle
$_0$ = orbital PE; $_1$ = force-centre spin KE; $_2$ = total; $_3$ = sub-satellite KE
Refer to chapter 14.3 for temperatures.

Venus

$\rho_{ave} = 5242.664311$ kg/m^3

Venus' matter is similar to that when it was ejected during the last '*Big-Bang*'. Its density and 'Δ' value show it to be an iron-rich planet with an internal structure somewhere between Mercury and that of the earth. Whilst it generates greater internal activity (δE) than Mercury, it generates less per unit mass; 0.40774 J/kg.

Because its internal frictional heat comes from E_0 & E_1 only (it has no satellites of its own) it has relatively low core-mantle activity, resulting in insignificant internal frictional heat or magnetic field. Because E_1 is greater than E_0, it will spin in the opposite direction to Mercury.

Venus has very little surface activity (volcanoes and earthquakes), due to the minimal internal activity, so its appearance will change very slowly during its existence as a planet. Moreover, its minimal internal activity will release little of its internal molecules, e.g. hydrocarbons and carbonyls; its atmosphere will have little CO_2. What there is of surface activity in Venus, is insufficient to generate continental drift but sufficient to evaporate its water, significantly increasing its atmospheric density.

Venus receives 172K of its surface temperature from the sun (1.911 times more than that received by the earth), which is more than sufficient to evaporate its surface water. The remainder of its surface heat comes from internal spin friction, little of which can escape Venus' dense H_2O atmosphere. Its highest surface temperature is said to be 734K.

Like Mercury, there is little to distinguish Venus from, say, Pluto other than its proximity to its force-centre, which is the reason Venus has acquired no satellites of its own.

14.6 Earth

Sym	Value	Units	Sym	Value	Units
Orbital Dimensions			**Core-Mantle Dimensions**		
t	*31558118.4*	s	rc	1215000	m
R^P	*1.470950E+11*	m	r	6371000.685	m
R^A	1.520942E+11	m			
Orbital Shape			**Planetary Spin**		
a	1.495945981E+11	m	Δ	0.334277698	
e	0.01670914665		J	1.082095E+37	kg.m²
b	1.495737135E+11	m	ω_0	1.9909886E-07	c/s
p	1.495528319E+11	m	E_0	2.144732E+23	J
f	1.47095E+11	m	ω_1	2.435001E-07	c/s
x'	2.499598078E+09	m	E_1	3.207996E+23	J
A	7.029445371E+22	m²	ω_2	7.292115E-05	c/s
L	9.398649712E+11	m	E_2	2.877018E+28	J
K	2.974914364E-19	s²/m³	ω_3	-7.292102E-05	c/s
v^P	3.028600879E+04	m/s	E_3	-2.877008E+28	J
g^P	-6.133232761E-03	m/s²			
v^A	2.929053557E+04	m/s			
g^A	-5.736671536E-03	m/s²			
Masses					
m_1	1.9885E+30	kg	mc	5.912786E+22	kg
m_2	*5.964520E+24*	kg	m_m	5.905392E+24	kg
Satellite Performance			**Core-Mantle Spin**		
F^P	-3.658178805E+22	N	Δ_c	1	
PE^P	-5.380998113E+33	J	Δ_m	0.341675424	
KE^P	2.735455000E+33	J	J_c	3.491441E+34	kg.m²
F^A	-3.421649078E+22	N	J_m	1.078604E+37	kg.m²
PE^A	-5.204129660E+33	J	Ω_c	3.505090E-06	c/s
KE^A	2.558586547E+33	J	E_c	2.144732E+23	J
E	-2.645543113E+33	J	ω_m	-7.303935E-05	c/s
h	4.454920463E+15	m²/s	E_m	-2.877040E+28	J
			$\delta\omega$	-6.953426E-05	c/s
			δE	-2.877061E+28	J
Core Pressure @ <1K					
pm	1.750E+24	N/m²	pe	2.183E+29	N/m²
Calculated Temperatures					
			T_r	90	K
			T_c	6000	K
			T_s	210	K
			T	300	K
Earth					
Input Data; Estimate					
ᴾ = perigee; ᴬ = Apogee; c = core; m = mantle					
0 = orbital PE; 1 = force-centre spin KE; 2 = total; s = sub-satellite KE					
Refer to chapter 14.3 for temperatures.					

The Mathematical Laws of Natural Science

Earth

ρ_{ave} = 5506.351327 kg/m³

Unlike Venus, the earth has a substantial satellite driving its angular motion, contrary to which, the sun is trying to drive its core in the opposite direction. Therefore, the earth's mantle and its core are revolving at different rates, generating internal friction and, as a result, internal heat. Earth generates 4823.746682 J/kg in its core, making it the third highest per unit mass in our solar system after Jupiter and Neptune. This activity is sufficient to melt its mantle matter but insufficient to melt its crust.

The hot core and [relatively] cold surface of the planet's matter drive its mantle plumes that rise and sink as temperatures change. These plumes drive the tectonic plates in its crust; continental drift, which is responsible for the planet's atmosphere and its weather. The moon's tilted orbital plane gives the earth its seasons. Along with the earth's proximity to its sun, all this internal and surface activity provides an ideal environment for life to flourish.

Life was created on a planet many universal periods ago. Its proteins have since spread throughout countless re-accretions and 'Big-Bangs'. It exists in almost all universal matter today.

Spin and core-pressure analysis shows us that the earth's mantle density just below its crust is not much more than that of liquid water, allowing heavier material, such as mountain roots, to fall from the crust.

Before acquiring its satellite, the earth was a cold celestial body, with a surface temperature of not much more than 90K. It was generating less internal energy than Venus; E_δ=-3.204358E+23 J due to its further proximity from its force-centre.

Less than 0.04% of earth's atmosphere is generated by its mantle activity, all the rest (N_2, O_2 & Ar) has been created by its surface life and potassium decay. 30% of the earth's atmospheric temperature is provided by the sun's radiated energy, the remainder (70%) is generated by its internal spin-friction.

Earth receives 90K of its surface temperature from the sun, and generates 210K from spin-friction. Its surface temperature is 300K.

The Mathematical Laws of Natural Science

14.7 Mars

Sym	Value	Units	Sym	Value	Units
Orbital Dimensions			**Core-Mantle Dimensions**		
t	*59354294.4*	s	r_c	461831.5834	m
R^P	*2.066550E+11*	m	r	3389500	m
R^A	2.492139E+11	m			
Orbital Shape			**Planetary Spin**		
a	2.279344353E+11	m	Δ	0.002317087	
e	0.09335770306		J	1.583269E+31	kg.m²
b	2.269389619E+11	m	ω_0	1.0585898E-07	c/s
p	2.259478361E+11	m	E_0	8.871154E+16	J
f	2.066550000E+11	m	ω_1	4.433650E-05	c/s
x'	2.127943533E+10	m	E_1	1.556136E+22	J
A	1.625058045E+23	m²	ω_2	7.088236E-05	c/s
L	1.429028790E+12	m	E_2	3.977416E+22	J
K	2.974914364E-19	s²/m³	ω_3	-5.530457E-05	c/s
v^P	2.649725040E+04	m/s	E_3	-2.421289E+22	J
g^P	-3.107373324E-03	m/s²			
v^A	2.197224924E+04	m/s			
g^A	-2.136686702E-03	m/s²			
Masses					
m_1	1.9885E+30	kg	m_c	3.247239E+21	kg
m_2	*6.417100E+23*	kg	m_m	N/A	kg
Satellite Performance			**Core-Mantle Spin**		
F^P	-1.994032536E+21	N	Δ_c	1	
PE^P	-4.120767937E+32	J	Δ_m	N/A	
KE^P	2.252736683E+32	J	J_c	2.770394E+32	kg.m²
F^A	-1.371133223E+21	N	J_m	-2.612067E+32	kg.m²
PE^A	-3.417054178E+32	J	Ω_c	N/A	c/s
KE^A	1.549022924E+32	J	E_c	N/A	J
E	-1.868031254E+32	J	ω_m	N/A	c/s
h	5.475789281E+15	m²/s	E_m	N/A	J
			$\delta\omega$	N/A	c/s
			δE	N/A	J
Core Pressure @ <1K					
pm	0	N/m²	pe	0	N/m²
Calculated Temperatures					
			T_r	38.7	K
			T_c	?	K
			T_s	188.23	K
			T	(227)	K

Mars
Input Data; Estimate
^{P} = perigee; ^{A} = Apogee; $_c$ = core; $_m$ = mantle
$_0$ = orbital PE; $_1$ = force-centre spin KE; $_2$ = total; $_3$ = sub-satellite KE
Refer to chapter 14.3 for temperatures.

Mars

$\rho_{ave} = 3934.080869$ kg/m³

Mars' matter is not similar to that when it was ejected during the last '*Big-Bang*'. Its density and 'Δ' value appear to show it to be a very different planet to the other three inner planets in our solar system. However, it's surface colour and structure may explain why;

1) It contains a great deal of oxidised iron.
2) During its early activity, sufficient internal heat must have been generated to maintain Mars' surface water in liquid form despite its distance from the sun. Its red surface colour (rust) indicates that it had accommodated oxygen-emitting plant life, perhaps earlier than the earth.
3) It has collected two small satellites that together manage to generate similar (>97%) planetary spin as that in the earth due to their much smaller orbits; 2.57% (Phobos) and 6.52% (Deimos) of Earth's lunar orbit.
4) It is a tiny planet (11% of earth's mass) with the solar system's largest volcano.

From the above information we may postulate that Mars originally had a density similar to the other inner planets in our solar system (≈5350 kg/m³), and orbited cold (≈39 K) and alone. It later trapped its first moon Phobos, which generated sufficient internal heat to turn Mars into a miniature Earth, giving life an opportunity to proliferate and liquid water to create its surface canals.

At some time later still, Mars trapped a second moon, the increased internal activity from which blew out a great deal of mantle matter onto the planet's surface, and in the process, created the largest volcano in the solar system.

Today the surface temperature of Mars is said to be 227K. Calculations show this to have been 290K before trapping Deimos, which is close to that of the earth (300K) and supports the probability of life previously.

Today, much of the planet's mantle is now missing (deposited onto its surface), which accounts for its lower-than-expected density, and is therefore responsible for a much-reduced internal energy which is why its surface water is frozen

The Mathematical Laws of Natural Science

Apart from the earth, Mars is by far the most interesting planet in our solar system, because *if* the following are true:

1) it is a hollow iron planet,

2) it had liquid water on its surface,

3) it has hosted oxygen-emitting plant-life,

4) its water has found its way into the planet's interior,

... it is possible that the kinetic energy in Deimos is keeping Mars' internal water liquid, which would mean that:

a) it may contain an atmosphere (of some description) internally,

b) it may contain life internally, and

c) it may be possible to access its internal voids.

Could it be that there is an atmosphere and water inside Mars, being kept liquid by an over-active moon?

14.8 Asteroid Belt

$\rho_{ave} = 2090$ kg/m³

It appears that there was once a planet at position '5' in our solar system and that it was impacted by a galactic comet, and is probably the source of many of our meteorites today and in the past. Perhaps this impact created the comet that hit the earth 63mn years ago; i.e. the asteroid belt was created 63mn years ago.

One of the reasons we know that our sun and its planets did not accrete from rocky particles is that the Asteroid belt remains a loose collection of rocks. I.e. orbiting celestial bodies do not accrete through *gravity*.

Because the moons in our solar system are trapped from galactic travel, it would appear from the Asteroid incident, that galactic comets can be quite substantial. However, whilst our own comets tend to orbit in cycles of hundreds of years, galactic comets will orbit in millions of years. But keep your eyes open!

The Mathematical Laws of Natural Science

14.9 Jupiter

Sym	Value	Units	Sym	Value	Units
Orbital Dimensions			**Core-Mantle Dimensions**		
t	374335689.6	s	r_c	3211180.751	m
R^P	7.405200E+11	m	r	69911000	m
R^A	8.156104E+11	m			
Orbital Shape			**Planetary Spin**		
a	7.780652166E+11	m	Δ	0.02278067	
e	0.04825458812		J	1.925855E+39	kg.m²
b	7.771588241E+11	m	ω_0	1.6784895E-08	c/s
p	7.762534876E+11	m	E_0	2.712882E+23	J
f	7.405200000E+11	m	ω_1	5.124219E-07	c/s
x'	3.754521656E+10	m	E_1	2.528419E+26	J
A	1.899659027E+24	m²	ω_2	1.758525E-04	c/s
L	4.885880874E+12	m	E_2	2.977768E+31	J
K	2.974914364E-19	s²/m³	ω_3	-1.758518E-04	c/s
v^P	1.370590205E+04	m/s	E_3	-2.977742E+31	J
g^P	-2.419979478E-04	m/s²			
v^A	1.244404703E+04	m/s			
g^A	-1.994893535E-04	m/s²			
Masses					
m_1	1.9885E+30	kg	m_c	1.091585E+24	
m_2	1.898190E+27	kg	m_m	1.897098E+27	
Satellite Performance			**Core-Mantle Spin**		
F^P	-4.593580846E+23	N	Δ_c	1	
PE^P	-3.401638488E+35	J	Δ_m	0.022784618	
KE^P	1.782891576E+35	J	J_c	4.502431E+36	kg.m²
F^A	-3.786686960E+23	N	J_m	1.921353E+39	kg.m²
PE^A	-3.088461391E+35	J	Ω_c	3.471418E-07	c/s
KE^A	1.469714479E+35	J	E_c	2.712882E+23	J
E	-1.618746912E+35	J	ω_m	-1.760584E-04	c/s
h	1.014949459E+16	m²/s	E_m	-2.977768E+31	J
			$\delta\omega$	-1.757113E-04	c/s
			δE	-2.977768E+31	J
Core Pressure @ <1K					
pm	5.570E+26	N/m²	pe	7.276E+29	N/m²
Calculated Temperatures					
			T_r	3.33	K
			T_c	15152.106	K
			T_s	1757.241	K
			T	1760.57	K

Jupiter
Input Data; Estimate
^{P} = perigee; ^{A} = Apogee; $_c$ = core; $_m$ = mantle
$_0$ = orbital PE; $_1$ = force-centre spin KE; $_2$ = total; $_3$ = sub-satellite KE
Refer to chapter 14.3 for temperatures.

The Mathematical Laws of Natural Science

Jupiter

$\rho_{ave} = 1326.216812$ kg/m³

Jupiter's matter is similar to that when it was ejected during the last '*Big-Bang*', but it is now a gas planet, which means it has collected sufficient satellite mass to generate the heat required to melt its crust. But what we see is mostly the planet's gaseous atmosphere; water plus the heavier hydrocarbon and silicon molecules in the planet's rock.

The above density quoted for this planet is incorrect, because it includes its gas cloud in the calculation. It is highly likely that there is little difference in this planet's viscous density to that of the inner iron-rich planets (≈ 5400 kg/m³). The planet's surface radius is expected to be 43781009.44 m; 6.876 times that of the earth.

Jupiter has collected a substantial satellite population (>50) - one of which (Ganymede) is twice the size of earth's moon - driving its angular motion, contrary to which, the sun is trying to drive its core in the opposite direction. Therefore, Jupiter's mantle and its core are revolving at different rates, generating internal friction and, as a result, internal heat. It generates the highest internal frictional energy per unit mass of any other body in our solar system; 15687.40748 J/kg in its core, apart from our sun (80513.156 J/kg).

Jupiter's weather is created by its moons orbiting in both directions (prograde and retrograde) that together generate sufficient competing kinetic energies in the planet's surface gases to account for its violent weather. The Red-Spot rotates because of its two adjacent gaseous layers, and molten mantle matter, rotating in opposite directions, which is caused by the opposing orbital directions of its moons.

As stated above, Jupiter's density and surface radius are both incorrect, because they include the gas cloud in their calculations. In the real world, Jupiter's internal energy is dependent upon realistic values for 'Δ' and angular velocity (ω), plus the revised density and surface radius above.

We can legitimately assume that Jupiter's 'Δ' value will be slightly less than that of the earth, because its core matter is much more mobile. But whilst the planet's angular velocity is unknown, hidden below its gas cloud, it can be safely assumed to be rotating considerably slower than the gas cloud. However, we have no idea what it is.

The Mathematical Laws of Natural Science

14.10 Saturn

Sym	Value	Units	Sym	Value	Units
Orbital Dimensions			**Core-Mantle Dimensions**		
t	929596608	s	r_c	1385801.125	m
R^P	1.352550E+12	m	r	58232000	m
R^A	1.501105E+12	m			
Orbital Shape			**Planetary Spin**		
a	1.426827696E+12	m	Δ	0.01406001	
e	0.05205792952		J	1.523923E+38	kg.m²
b	1.424893012E+12	m	ω_0	6.759045 0E-09	c/s
p	1.422960953E+12	m	E_0	3.480997E+21	J
f	1.352550000E+12	m	ω_1	3.473221E-07	c/s
x'	7.427769562E+10	m	E_1	9.191740E+24	J
A	6.387099181E+24	m²	ω_2	1.637867E-04	c/s
L	8.958945951E+12	m	E_2	2.044045E+30	J
K	2.974914364E-19	s²/m³	ω_3	-1.637864E-04	c/s
v^P	1.015981492E+04	m/s	E_3	-2.044036E+30	J
g^P	-7.254017694E-05	m/s²			
v^A	9.154359018E+03	m/s			
g^A	-5.889289434E-05	m/s²			
Masses					
m_1	1.9885E+30	kg	m_c	8.773362E+22	
m_2	5.683400E+26	kg	m_m	5.682523E+26	
Satellite Performance			**Core-Mantle Spin**		
F^P	-4.122748416E+22	N	Δ_c	1	
PE^P	-5.576223370E+34	J	Δ_m	0.014061969	
KE^P	2.933255007E+34	J	J_c	6.739503E+34	kg.m²
F^A	-3.347118757E+22	N	J_m	1.523249E+38	kg.m²
PE^A	-5.024378011E+34	J	Ω_c	3.214052E-07	c/s
KE^A	2.381409647E+34	J	E_c	3.480997E+21	J
E	-2.642968364E+34	J	ω_m	-1.638230E-04	c/s
h	1.374165767E+16	m²/s	E_m	-2.044045E+30	J
			$\delta\omega$	-1.635016E-04	c/s
			δE	-2.044045E+30	J
Core Pressure @ <1K					
pm	1.668E+26	N/m²	pe	5.457E+29	N/m²
Calculated Temperatures					
			T_r	0.99	K
			T_c	14099.30	K
			T_s	1492.327	K
			T	1493.32	K

Saturn
Input Data; Estimate
^{P} = perigee; ^{A} = Apogee; $_c$ = core; $_m$ = mantle
$_0$ = orbital PE; $_1$ = force-centre spin KE; $_2$ = total; $_3$ = sub-satellite KE
Refer to chapter 14.3 for temperatures.

Saturn

$\rho_{ave} = 687.1230137$ kg/m³

Saturn's matter is similar to that when it was ejected during the last '*Big-Bang*', but it is now a gas planet, which means it has collected sufficient satellite mass to generate the heat required to melt its crust. But what we see is mostly the planet's gaseous atmosphere; the lighter water, hydrocarbon and silicon molecules in the planet's rock.

The above density quoted for this planet is incorrect, because it includes its gas cloud in the calculation. It is highly likely that there is little difference in this planet's viscous density to that of the inner iron-rich planets (≈ 5400 kg/m³). The planet's surface radius is expected to be 29289274.75m; 4.6 times that of the earth.

Saturn has collected a substantial satellite population (>50) driving its angular motion. Whilst Saturn is only 30% of Jupiter's *mass*, its largest moon (Titan) is almost 92% as massive as Jupiter's largest moon, therefore Saturn may be more active [internally] than Jupiter.

Saturn generates less internal frictional energy per unit mass (3596.592 J/kg) than the earth, but achieves a surface temperature not much less than that of Jupiter (1493.32) despite its greater distance from the sun's radiated heat. This anomaly is due to Saturn generating one tenth the spin energy of Jupiter, but 100 times that of the earth.

The low *apparent* density (including its gas cloud) simply tells us that Saturn's atmospheric gases are lighter than those surrounding Jupiter.

Saturn's rings are most probably the remains of a satellite that was pulled apart by the huge potential forces induced by orbiting between such a large planet and a substantial moon. If the satellite was an ice moon, it would not have required much kinetic and potential energy to pull it apart.

Saturn and Jupiter have collected the most moons because they are the most massive planetary satellites in our solar system and sufficiently distant from their force-centre's gravitational influence.

As stated above, Saturn's density and surface radius are both incorrect, because they include their gas cloud in their calculations.

The Mathematical Laws of Natural Science

In the real world, Saturn's internal energy is dependent upon realistic values for 'Δ' and angular velocity (ω), plus the revised density and surface radius that have been provided above.

We can legitimately assume that Saturn's 'Δ' value will be slightly less than that of the earth, because its core matter is much more mobile. But whilst the planet's angular velocity is unknown, hidden below its gas cloud, it can be safely assumed to be rotating considerably slower than the gas cloud. However, we have no idea what it is.

The Mathematical Laws of Natural Science

The Mathematical Laws of Natural Science

14.11 Uranus

Sym	Value	Units	Sym	Value	Units
Orbital Dimensions			**Core-Mantle Dimensions**		
t	2651218560	s	r_c	1115922.071	m
R^P	2.741300E+12	m	r	25362000	m
R^A	2.997691E+12	m			
Orbital Shape			**Planetary Spin**		
a	2.869495390E+12	m	Δ	0.02493762	
e	0.04467523818		J	1.389062E+37	kg.m²
b	2.866630380E+12	m	ω_0	2.3699236E-09	c/s
p	2.863768230E+12	m	E_0	3.900860E+19	J
f	2.741300000E+12	m	ω_1	6.358382E-08	c/s
x'	1.281953900E+11	m	E_1	2.807922E+22	J
A	2.584205837E+25	m²	ω_2	-1.012377E-04	c/s
L	1.802057180E+13	m	E_2	-7.118296E+28	J
K	2.974914364E-19	s²/m³	ω_3	1.012377E-04	c/s
v^P	7.111398264E+03	m/s	E_3	7.118298E+28	J
g^P	-1.765924516E-05	m/s²			
v^A	6.503164433E+03	m/s			
g^A	-1.476765724E-05	m/s²			
Masses					
m_1	1.9885E+30	kg	m_c	4.581049E+22	
m_2	8.681300E+25	kg	m_m	8.676719E+25	
Satellite Performance			**Core-Mantle Spin**		
F^P	-1.533052050E+21	N	Δ_c	1	
PE^P	-4.202555585E+33	J	Δ_m	0.024947870	
KE^P	2.195152879E+33	J	J_c	2.281879E+34	kg.m²
F^A	-1.282024628E+21	N	J_m	1.386780E+37	kg.m²
PE^A	-3.843113407E+33	J	Ω_c	5.847212E-08	c/s
KE^A	1.835710700E+33	J	E_c	3.900860E+19	J
E	-2.007402707E+33	J	ω_m	1.013209E-04	c/s
h	1.949447606E+16	m²/s	E_m	7.118296E+28	J
			$\delta\omega$	1.013794E-04	c/s
			δE	7.118296E+28	J
Core Pressure @ <1K					
pm	2.548E+25	N/m²	pe	3.638E+29	N/m²
Calculated Temperatures					
			T_r	0.24	K
			T_c	8742.325	K
			T_s	610.667	K
			T	610.91	K

Uranus
Input Data; Estimate
^{P} = perigee; ^{A} = Apogee; $_c$ = core; $_m$ = mantle
$_0$ = orbital PE; $_1$ = force-centre spin KE; $_2$ = total; $_3$ = sub-satellite KE
Refer to chapter 14.3 for temperatures.

The Mathematical Laws of Natural Science

Uranus

$\rho_{ave} = 1270.415139$ kg/m³

Uranus's matter is similar to that when it was ejected during the last '*Big-Bang*', but it is now a gas planet, which means it has collected sufficient satellite mass to generate the heat required to melt its crust. But what we see is mostly the planet's gaseous atmosphere; the lighter elements plus water and silicon molecules in the planet's matter.

The above density quoted for this planet is incorrect, because it includes its gas cloud in the calculation. It is highly likely that there is little difference in this planet's viscous density to that of the inner iron-rich planets (≈ 5400 kg/m³). The planet's surface radius is expected to be 15656720.49m; 2.45 times that of the earth.

Uranus has collected a substantial satellite population (≈ 30) driving its angular motion.

Whilst Uranus generates less internal frictional energy per unit mass (819.9575 J/kg) than the earth, and despite being so much further from stellar radiation, it achieves a higher surface temperature (610.6666 K). The planet's *apparent* density (including its gas cloud) simply tells us that Uranus's atmospheric gases are similar to those created by Jupiter; giving Uranus its colour.

As stated above, Uranus's density and surface radius are both incorrect, because they include the gas cloud in their calculations.

In the real world, Uranus's internal energy is dependent upon realistic values for 'Δ' and angular velocity (ω), plus the revised density and surface radius that have been provided above.

We can legitimately assume that Uranus's 'Δ' value will be slightly less than that of the earth, because its core matter is much more mobile. But whilst the planet's angular velocity is unknown, hidden below its gas cloud, it can be safely assumed to be rotating considerably slower than the gas cloud. However, we have no idea what it is.

The Mathematical Laws of Natural Science

14.12 Neptune

Sym	Value	Units	Sym	Value	Units
Orbital Dimensions			**Core-Mantle Dimensions**		
t	5200329600	s	r_c	1396768.826	m
R^P	4.444450E+12	m	r	24622000	m
R^A	4.548298E+12	m			
Orbital Shape			**Planetary Spin**		
a	4.496373960E+12	m	Δ	0.06523793	
e	0.01154796298		J	1.056968E+38	kg.m²
b	4.496074142E+12	m	ω_0	1.2082283E-09	c/s
p	4.495774344E+12	m	E_0	7.714895E+19	J
f	4.444450000E+12	m	ω_1	6.294054E-09	c/s
x'	5.192396004E+10	m	E_1	2.093597E+21	J
A	6.351053352E+25	m²	ω_2	1.083383E-04	c/s
L	2.825060890E+13	m	E_2	6.202913E+29	J
K	2.974914364E-19	s²/m³	ω_3	-1.083382E-04	c/s
v^P	5.495748646E+03	m/s	E_3	-6.202913E+29	J
g^P	-6.718142608E-06	m/s²			
v^A	5.370268285E+03	m/s			
g^A	-6.414864102E-06	m/s²			
Masses					
m_1	1.9885E+30	kg	m_c	8.983321E+22	
m_2	1.024130E+26	kg	m_m	1.023232E+26	
Satellite Performance			**Core-Mantle Spin**		
F^P	-6.880251390E+20	N	Δ_c	1	
PE^P	-3.057893329E+33	J	Δ_m	0.065350148	
KE^P	1.546602884E+33	J	J_c	7.010452E+34	kg.m²
F^A	-6.569654772E+20	N	J_m	1.056267E+38	kg.m²
PE^A	-2.988074714E+33	J	Ω_c	4.691449E-08	c/s
KE^A	1.476784269E+33	J	E_c	7.714895E+19	J
E	-1.511290445E+33	J	ω_m	-1.083742E-04	c/s
h	2.442558007E+16	m²/s	E_m	6.119540E+29	J
			$\delta\omega$	-1.083273E-04	c/s
			δE	-6.202913E+29	J
Core Pressure @ <1K					
pm	3.005E+25	N/m²	pe	4.365E+29	N/m²
Calculated Temperatures					
			T_r	0.1	K
			T_c	9341.467	K
			T_s	642.928	K
			T	643.03	K

Neptune
Input Data; Estimate
^{P} = perigee; ^{A} = Apogee; $_c$ = core; $_m$ = mantle
$_0$ = orbital PE; $_1$ = force-centre spin KE; $_2$ = total; $_3$ = sub-satellite KE
Refer to chapter 14.3 for temperatures.

Neptune

$\rho_{ave} = 1637.934377$ kg/m³

Neptune's matter is similar to that when it was ejected during the last '*Big-Bang*', but it is now a gas planet, which means it has collected sufficient satellite mass to generate the heat required to melt its crust. But what we see is mostly the planet's gaseous atmosphere; the lighter elements plus water and silicon molecules in the planet's matter.

The above density quoted for this planet is incorrect, because it includes its gas cloud in the calculation. It is highly likely that there is little difference in this planet's viscous density to that of the inner iron-rich planets (≈ 5400 kg/m³). The planet's surface radius is expected to be 16543379.78 m; 2.6 times that of the earth.

Whilst Neptune has collected a substantial satellite population (≈ 15) driving its angular motion, its largest moon (Triton) orbits in the opposite direction to most of the others. It is expected therefore that Neptune's surface matter will suffer significant levels of disruption.

Neptune generates more internal frictional energy per unit mass (6056.7633 J/kg) than the earth, and despite being so much further from stellar radiation, it achieves a higher surface temperature (643.03 K). The planet's *apparent* density (including its gas cloud) simply tells us that Neptune's atmospheric gases are actually heavier than those created by Jupiter.

As stated above, Neptune's density and surface radius are both incorrect, because they include the gas cloud in their calculations.

In the real world, Neptune's internal energy is dependent upon realistic values for 'Δ' and angular velocity (ω), plus the revised density and surface radius that have been provided above.

We can legitimately assume that Neptune's 'Δ' value will be slightly less than that of the earth, because its core matter is much more mobile. But whilst the planet's angular velocity is unknown, hidden below its gas cloud, it can be safely assumed to be rotating considerably slower than the gas cloud. However, we have no idea what it is.

The Mathematical Laws of Natural Science

14.13 Pluto

Sym	Value	Units	Sym	Value	Units
Orbital Dimensions			**Core-Mantle Dimensions**		
t	7824384000	s	r_c	76464.36337	m
R^P	4.436820E+12	m	r	1187000	m
R^A	7.371060E+12	m			
Orbital Shape			**Planetary Spin**		
a	5.903940173E+12	m	Δ	8.64242	
e	0.02484984824		J	5.484999E+35	kg.m²
b	5.718747063E+12	m	ω_0	8.0302619E-10	c/s
p	5.539363037E+12	m	E_0	1.768504E+17	J
f	4.436820000E+12	m	ω_1	1.512158E-10	c/s
x'	1.467120173E+12	m	E_1	6.271063E+15	J
A	1.060700342E+26	m²	ω_2	-1.138559E-05	c/s
L	3.651627701E+13	m	E_2	-3.555149E+25	J
K	2.974914364E-19	s²/m³	ω_3	1.138559E-05	c/s
v^P	6.110837548E+03	m/s	E_3	3.555149E+25	J
g^P	-6.741268858E-06	m/s²			
v^A	3.678261332E+03	m/s			
g^A	-2.442448046E-06	m/s²			
Masses					
m_1	1.9885E+30	kg	m_c	1.473807E+19	
m_2	1.303E+22	kg	m_m	1.301526E+22	
Satellite Performance			**Core-Mantle Spin**		
F^P	-8.783873322E+16	N	Δ_c	1	
PE^P	-3.897246483E+29	J	Δ_m	8.665309303	
KE^P	2.432853160E+29	J	J_c	3.446821E+28	kg.m²
F^A	-3.182509804E+16	N	J_m	5.484999E+35	kg.m²
PE^A	-2.345847182E+29	J	Ω_c	3.203381E-06	c/s
KE^A	8.814538586E+28	J	E_c	1.768504E+17	J
E	-1.464393323E+29	J	ω_m	1.138559E-05	c/s
h	2.711268625E+16	m²/s	E_m	3.555149E+25	J
			$\delta\omega$	1.458897E-05	c/s
			δE	3.555149E+25	J
Core Pressure @ <1K					
pm	3.824E+21	N/m²	pe	7.646E+27	N/m²
Calculated Temperatures					
			T_r	0.06	K
			T_c	<20K	K
			T_s	19	K
			T	19	K

Pluto
Input Data; Estimate
^{P} = perigee; ^{A} = Apogee; $_c$ = core; $_m$ = mantle
$_0$ = orbital PE; $_1$ = force-centre spin KE; $_2$ = total; $_3$ = sub-satellite KE
Refer to chapter 14.3 for temperatures.

Pluto

$\rho_{ave} = 1859.960193$ kg/m³

Pluto is a small ice planet in which all of its matter exists in solid form, hence its relatively high density.

Having acquired a substantial sub-satellite population, the largest of which (Charon) is more than 10% as massive as the planet itself, the competing kinetic and potential energies are pulling Pluto into a localised orbit, and is the reason it has a 'Δ' value greater than 1.

Pluto generates more than half as much spin energy as the earth from its five moons (2728.4336 J/kg), but because it is so small, it is being pulled into a local orbit and cannot therefore generate internal frictional heat between its core and its mantle. Its surface temperature will not be much more than 20K.

Despite its ludicrous relegation from planetary status; Pluto is more entitled to be called a planet than either Mercury or Venus. Whilst all three are solid lumps of matter, Pluto is only a lump of rock because it has collected so much lunar mass that it is being pulled into local orbit, preventing the sun's potential energy from acting at the planet's core. I.e. there is no conflicting core-mantle spin to generate internal heat (energy), otherwise, Pluto would actually be a gas-planet

Whilst there are planets outside Pluto's orbit; MakeMake, Haumea, Eris, etc., as we know very little about them, I have not addressed them here.

The Mathematical Laws of Natural Science

15 The Physical Constants

Because everything in the universe is energy, that are defined using only [electrical or magnetic] charge, distance and time, every constant included in this chapter has been defined using *only* these four constants (metric units):

Electricity: Coulomb {C}

Mass: kilogram {kg}

Length: metre {m}

Time: second {s}

Temperature is not a genuine physical variable, but it may be used for the sake of convenience:

Temperature: Kelvin {K}

Notes:

1) Joules and Newtons remain useful, but they are merely compilations of the above.
2) When converting to imperial units ...
... between numerators or denominators: multiply by the conversion factor above,
... across numerators and denominators: divide by the conversion factor above.
All new constants (unknown until now) are highlighted in **bold** *text.*

In the following tables, all of the constants are <u>exact</u> to the number of decimals stated; there are no approximations, tolerances or estimates.

Whilst *mass* is actually *magnetic charge*, I shall continue to refer to it as *mass* in order to prevent confusion.

Whilst *gravity* is actually *magnetism*, I shall continue to refer to it as *gravity* in order to prevent confusion.

The Mathematical Laws of Natural Science

15.1 Primary Constants

All universal constants, both old and new, can be defined from only; *four* primary constants, *two* ratios and a particle constant (Σ), all of which are listed below:

Symbol	Value	Units
m_e	**9.1093897E-31**	*kg*
mass of an electron (elementary magnetic charge unit)		
e	1.60217648753E-19	C
elementary electrical charge unit		
R_n	2.81793795383896E-15	m
neutronic radius		
t_n	5.90596121302193E-23	s
neutronic period		
ξ_m	1836.15115053207	
static ratio		
ξ_v	1722.0458764934	
dynamic ratio		
Σ	3E-91 (exact)	m^6
particle constant		

Table 15.1

The Mathematical Laws of Natural Science

15.1.1 Principal Constants

These are the most important *non-primary* constants.

Symbol	Formula	Value	Units
G	$a_o.c^2 / m_u$	6.67359232004334E-11	$m^3 / s^2.kg$
Newton's gravitational constant			chapter 15.5.3
k	$1/\varepsilon_o$	8.98755184732667E+09	$J.m / C^2$
Coulomb's constant			chapter 15.5.5
k'	$k.RC^2$	2.78024810626745E+32	$m^3 / kg.s^2$
Coulomb's constant (*modified*)			chapter 15.5.5
φ	$G.m_e.m_p / k.e^2$	4.40742111792334E-40	
coupling ratio			chapter 15.5.4
μ_o	$R_a.m_e/e^2$	1E-07	$kg.m / C^2$
Henry's magnetic constant			chapter 15.5.5
μ	$4\pi.\mu$	1.25663706143592E-06	$kg.m / C^2$
magnetic constant (*spherical*)			chapter 15.5.5
ε_o	$1/\mu_o.c^2$	1.11265004863082E-10	$C^2 / J.m$
permittivity constant			chapter 15.5.5
ε	$1/\mu.c^2$	8.85418775855161E-12	$C^2 / J.m$
permittivity constant			chapter 15.5.5
h	$\tfrac{1}{2}.m_e.c.\xi_v . R_a$	6.62607174469163E-34	$kg.m^2/s$
Planck's constant			chapter 15.5.6
h'	$\tfrac{1}{2}.m_e.c^2 . R_a$	1.15353857232684E-28	$J.m$
Planck's constant (*modified*)			chapter 15.5.6
$\hbar$	$h/2\pi$	1.054572071449210E-34	$kg.m^2/s$
Planck's constant (Dirac)			
γ	$(\xi_v / 4\pi)^2$	18778.8808461551	
Rydberg's constant			
a_o	$R_a.\gamma$	5.2917721067E-11	m
Rydberg's radius			chapter 15.5.8
R_∞	$1 / a_o.\xi_v$	1.09737269561359E+07	$/m$
Rydberg's wave number			chapter 15.5.8
R_Y	$\tfrac{1}{2}.m_e.c^2 / \gamma$	2.17987197684936E-18	J
Rydberg's electron energy constant			chapter 15.5.8
X	T_a/c^2	6.9353271647894E-09	$K.s^2/m^2$
heat transfer coefficient (*velocity*)			chapter 15.5.17
X_R	$T_a.R_a$	1.75646616508035E-06	$K.m$
heat transfer coefficient (*radial*)			chapter 15.5.17
Y	$\sqrt[3]{[\tfrac{1}{2}.\xi]}$	9.51345439232503	
temperature coefficient			
e_a	$m_p.RC$	2.94183820093364E-16	C
proton charge (*neutronic*)			
Table 15.1.1a			

The Mathematical Laws of Natural Science

These are the most important *non-primary* constants (continued).

Symbol	Formula	Value	Units
ρ_u	$m_u.\sqrt{[\xi_m/\Sigma]}$	7.1266079635045E+16	kg/m³
ultimate density			
m_u	$\rho_u/1$	7.1266079635045E+16	kg
unit mass of ultimate density			
R_x	$X_R/\underline{T}_x$	8.598540985 72228E-07	m
Cold orbital radius			chapter 2.5.12
R_o	$R_n.\xi_r^2$	8.3564315638157E-09	m
Planck minimum orbital radius			chapter 2.5.12
R_m	$R_n.\xi_r$	4.85261843362263E-12	m
Planck mean orbital radius			chapter 2.5.12
R_r	R_o/κ	8.40016460895157E-11	m
EME orbital radius constant			chapter 2.5.12
v_x	$\sqrt{[\underline{T}_x/X]}$	17162.2425219270	m/s
electron cold velocity			chapter 2.5.12
v_o	c/ξ_r	174090.866621084	m/s
electron minimum orbital velocity (Planck)			chapter 2.5.12
v_m	$c/\sqrt{\xi_r}$	7224342.80705004	m/s
electron mean orbital velocity (Planck)			chapter 2.5.12
c	$2\pi.R_n/t_n$	299792459	m/s
electron neutronic velocity			
$\underline{T}_x$	$X.(c/Y.\xi_m)^2$	2.04274907568265	K
cold temperature			chapter 2.5.12
$\underline{T}_o$	$X.v_o^2$	210.193328535837	K
Planck minimum temperature			chapter 2.5.12
$\underline{T}_m$	$X.v_m^2$	361962.554671561	K
Planck mean temperature			chapter 2.5.12
$\underline{T}_n$	$X.c^2$	623316124.717179	K
neutronic temperature			chapter 2.5.12
h_e	$c.R_n$	8.4479654849081E-07	m²/s
Newton's constant of motion (*electron*)			
κ	$(2\pi)2 \cdot 24/3$	99.4793787125405	
EME wavelength (constant)			
K	t_n^2/R_n^3	0.15587874533403	s²/m³
constant of proportionality (proton-electron pair)			chapter 6.1
e	$\exp(1)$	2.71828182845905	
Natural logarithm			
ϵ_e	m_e/ρ_u	7.82336489952175E+46	
ϵ_p	m_p/ρ_u	4.26074122343073E+43	
number of particles in a unit mass of ultimate density			
Unknown constant			
Table 15.1.1b			

The Mathematical Laws of Natural Science

15.2 Heat Capacities

Constants and formulas for heat generated by the proton-electron pair.

Symbol	Formula	Value	Units
RAM_e	$m_e . N_A$	5.4858031839070700E-07	kg/mol
Relative atomic mass of an electron			
RAM_p	$m_p . N_A$	1.00727638277235E-03	kg/mol
Relative atomic mass of a proton (also the RAM of an hydrogen atom)			
R_e	$R_i / RAM_e = k_B / m_e$	1.51563563034308E+07	J/kg.K
Specific gas constant for an electron			
R_p	$R_i / RAM_p = k_B / m_p$	8.25441647276088E+03	J/kg.K
Specific gas constant for a proton			
C_{et}	k_B / m_e	1.51563563034305E+07	J/kg.K
Specific heat capacity for the electron (constant temperature)			
C_{eV}	$1.5 . C_{et}$	2.27345344551458E+07	J/kg.K
Specific heat capacity for the electron (constant volume)			
C_{ep}	$C_{et} + C_{eV}$	3.78908907585763E+07	J/kg.K
Specific heat capacity for the electron (constant pressure)			
C_{pt}	k_B / m_p	8.25441647276074E+03	J/kg.K
Specific heat capacity for the proton (constant temperature)			
C_{pV}	$1.5 . C_{pt}$	1.23816247091411E+04	J/kg.K
Specific heat capacity for the proton (constant volume)			
C_{pp}	$C_{pt} + C_{pV}$	2.06360411819018E+04	J/kg.K
Specific heat capacity for the proton (constant pressure)			
C_t	$m_e . C_{et}$ $m_p . C_{pt}$	1.38065156E-23	J/K
Heat capacity (constant temperature); equal to R & Qt			
C_V	$m_e . C_{eV}$ $m_p . C_{pV}$	2.07097734E-23	J/K
Heat capacity (constant volume); equal to QV			
C_p	$m_e . C_{ep}$ $m_p . C_{pp}$	3.4516289E-23	J/K
Heat capacity (constant pressure); equal to Qp			

Table 15.2

Symbol	Formula	Value	Units
k_B	$m_e / X.Y$	1.38065156E-23	J/K
Boltzmann's constant			
N_A		6.02214129E+23	/mole
Avogadro's number (original)			
N_A	$2 / (m_e + m_p + m_n) . 1000$	5.97538412973187E+23	/mole
Avogadro's number (based upon C_{12})			
R_i	$k_B . N_A$	8.31447876657891	J/K.mol
Ideal gas constant; Avogadro's number (original)			
R_i	$k_B . N_A$	8.24992342031355	J/K.mol
Ideal gas constant; Avogadro's number (based upon C_{12})			

Table 15.2-1; The Energy Constants

The Mathematical Laws of Natural Science

15.3 Charge Capacities

Constants and formulas for electrical charge in the proton-electron pair.

Symbol	Formula	Value	Units
RC	e/m_e	1.75881869180545E+11	C/kg
Relative charge capacity			
RAC_e	$e.N_A$	96485.3317942158	C/mol
Relative atomic charge of an electron (also equal to the Farad)			
RAC_p	$e'.N_A$	1.77161652983418E+08	C/mol
Relative atomic charge of a proton (also the RAC of an hydrogen atom)			
R_{ee}	R_i / RAC_e	8.61735002820125E-05	J / C.K
Specific gas constant for an electron			
R_{ep}	R_i / RAC_p	4.69315939796359E-08	J / C.K
Specific gas constant for a proton			
q_{et}	k_B / m_e	8.61735002820123E-05	J / C.K
Specific charge capacity for the electron (constant temperature)			
q_{ev}	$1.5 \cdot q_{et}$	1.29260250423019E-04	J / C.K
Specific charge capacity for the electron (constant volume)			
q_{ep}	$q_{et} + q_{ev}$	2.15433750705 03E-04	J / C.K
Specific charge capacity for the electron (constant pressure)			
q_{pt}	k_B / e'	4.69315939796358E-08	J / C.K
Specific charge capacity for the proton (constant temperature)			
q_{pv}	$1.5 \cdot q_{pt}$	7.03973909694 54E-08	J / C.K
Specific charge capacity for the proton (constant volume)			
q_{pp}	$q_{pt} + q_{pv}$	1.1732898494909E-07	J / C.K
Specific charge capacity for the proton (constant pressure)			
Q_t	$e.q_{et}$ $e'.q_{pt}$	1.38065156E-23	J/K
Charge capacity (constant temperature); equal to R & Ct			
Q_V	$e.q_{ev}$ $e'.q_{pv}$	2.07097734E-23	J/K
Charge capacity (constant volume); equal to CV			
Q_p	$e.q_{ep}$ $e'.q_{pp}$	3.4516289E-23	J/K
Charge capacity (constant pressure); equal to Cp			
Q_p	$e.q_p$		J/K
Charge capacity (constant pressure); Q_t multiplied by 2.5; also equal to C_p			

Table 15.3

The Mathematical Laws of Natural Science

15.4 Unity

Today we have to deal with numerous disparate units that vary with country, language, and/or branch of science, the conversions between which make our lives so much more complicated than they need to be. For example; how many of us know what a *crocodile* is, or how to use it?

Why should the unit for distance be based upon the pendulum, or even a body part?

But given that all of nature is energy, and that it comprises the following;

$\quad$ magnetic; kg.(m/s)² & electrical; C.(m/s)²

and that every other constant in science is derived from just these units; life would be so much simpler if we set the primary constants to unity ...

Symbol	Value	Units
m_e	1	G
electron mass (magnetic charge)		
e	1	C
electron electrical charge		
R_n	1	m
neutronic radius		
t_n	1	s
neutronic period		
ξ_m	1836.15115053207	
static ratio		
ξ_v	1722.0458764934	
dynamic ratio		
Σ	3E-91	m^6
particle constant		
Table 15.4		

Because all of the following constants are derived purely from two ratios (plus π), the need for conversions simply vanish, all units are the same; neutronic. But this discovery also shows us that;

our universe comprises just two particles that are related by two ratios

The Mathematical Laws of Natural Science

15.4.1 Principal Constants

All of these constants are calculated using the unity primary values.

Symbol	Formula	Value	Units
G	$\varphi \cdot k/\xi_m$	9.47623573370968E-42	$m^3/s^2.G$
Newton's gravitational constant			
k	$(2\pi)^2$	39.4784176043574	$J.m/C^2$
Coulomb's constant [for an electron]			
k'	$k.RC^2$	39.4784176043574	$m^3/s^2.kg$
Coulomb's constant (*modified*)			
φ	$(\xi_r/4\pi)^2 \cdot \sqrt{[\sum . \xi_m]}$	4.40742111792334E-40	
Coupling Ratio			
μ_o	$R_n.m_e/e^2$	1	$G.m/C^2$
magnetic constant (fundamental)			
μ	4π	12.5663706143592	$G.m/C^2$
magnetic constant (spherical)			
ε_o	$1/(2\pi)^2$	2.53302959105844E-02	$C^2/J.m$
Permittivity of a vacuum (e.g. within an atom)			
ε	$1/2.(2\pi)^3$	2.01572090207497E-03	$C^2/J.m$
Permittivity of a vacuum (e.g. within an atom)			
h	$\pi.\xi_r$	5409.96667473626	$G.m^2/s$
Planck's constant (resolved into its component parts)			
h'	$2.\pi^2$	19.7392088021787	J.m
Modified Planck's constant			
$\hbar$	$h/2\pi$	861.0229382467	$G.m^2/s$
Plank's constant (Dirac)			
γ	$(\xi_v/4\pi)^2$	18778.8808461551	
Rydberg's constant			
a_o	γ	18778.8808461551	m
Rydberg's radius			
R_∞	$(4\pi)^2/\xi_r^3$	3.09232816847610E-08	/m
Rydberg's wave number			
R_Y	$2.(4.\pi^2/\xi_r)^2$	1.05113872141216E-03	J
Rydberg's universal constant for the energy of an electron			
X	$T_n/(2\pi)^2$	1.57887818849249E+07	$K.s^2/m^2$
heat transfer coefficient (velocity)			
X_R	T_n	6.23316124717178E+08	K.m
heat transfer coefficient (radial)			
Y	$\sqrt[3]{[½.\xi_r]}$	9.51345439232503	
Temperature coefficient			
e_n	ξ_m	1836.15115053207	C
Proton charge (neutronic)			

Table 15.4.1a

The Mathematical Laws of Natural Science

Symbol	Formula	Value	Units
ρ_u	$\sqrt{[\xi_m/\Sigma]}$	7.82336489952175E+46	G/m³
ultimate density			
m_u	$\sqrt{[\xi_m/\Sigma]}$	7.82336489952175E+46	G
unit mass of ultimate density			
R_x	T_x/X_R	8.59854098572228E-07	m
Cold orbital radius			
R_o	ξ_r^2	2.96544200074792E+06	m
Planck minimum orbital radius			
R_m	ξ_r	1722.0458764934	m
Planck mean orbital radius			
R_r	$\xi_r^2/$	29809.6152099721	m
EME orbital radius constant			
v_x	$\sqrt{[T_x/X]}$	5.18824113201723E-04	m/s
electron cold velocity			
v_o	$2\pi/\xi_r$	3.64867474957986E-03	m/s
electron minimum Planck orbital velocity			
v_m	$2\pi/\sqrt{\xi_r}$	0.15141102858523	m/s
electron mean Planck orbital velocity			
C	2π	6.28318530717959	m/s
electron neutronic velocity			
T_x	$X.(c/Y.\xi_m)^2$	2.04274907568265	K
cold temperature			
T_o	T_n/ξ_r^2	210.193328535837	K
Planck minimum temperature			
T_m	T_n/ξ_r	361962.554671561	K
Planck mean temperature			
T_n	$(2\pi)^2/Y.k_B$	623316124.717179	K
neutronic temperature			
h_e	2π	6.28318530717959	m²/s
Newton's constant of motion (*electron*)			
K	$24.(2\pi)^2/3$	99.4793787125405	
EME wavelength (constant)			
K		1	s²/m³
Constant of proportionality (proton-electron pair)			
E	exp(1)	2.71828182845905	
Natural logarithm			
ϵ_e	m_e/ρ_u	7.82336489952175E+46	
ϵ_p	m_p/ρ_u	4.26074122343073E+43	
number of particles in a unit mass of ultimate density			
Table 15.4.1b			

The Mathematical Laws of Natural Science

15.5 Physical Constants - Explained

All the physical constants we use today were derived by the Heroes listed in Appendix A-4 of this publication. The trouble is, very few of us understand their origins or their meaning, without which it becomes very difficult to apply them correctly or appropriately.

Most mathematical and scientific constants (chapters 15.1 to 3) are self-explanatory. These have not been explained here. But there are a number of new ones I have discovered during my work on atomic and celestial orbits. These constants are included here.

This chapter provides a detailed explanation of the most important constants that have been used in this work.

15.5.1 Σ

This universal constant is the reciprocal of the product of the atomic particles (protons and electrons) in a unit mass of ultimate density:

ultimate density: $\rho_u = m_e / (^4/_3\pi . r^3) = 7.12660796350045\text{E}+16$ kg/m³
Newtonian mass: $m_u = 7.12660796350045\text{E}+16$ kg
Number of electrons: $\epsilon_e = m_u / m_e$
Number of protons: $\epsilon_p = m_u / m_p$
$\Sigma = 1 / (\epsilon_e . \epsilon_p) = 3\text{E-}91$

Whilst the reasoning behind this bizarrely accurate value has yet to be discovered, it applies to all the following relationships, including the unification of Newton's and Plack's particles:

$G = \sqrt{[\Sigma . a_o^2 . c^4 / m_p . m_e]}$

$m_e . m_p = \Sigma . \rho_u^2$

$\rho_u = m_e . \sqrt{[\xi_m / \Sigma]}$

$V_e . V_p = \Sigma$
$V^{P2} = r^{P6} . (^4/_3 . \pi)^2 = \Sigma$

$r^P = \sqrt[6]{[\Sigma / (^4/_3 . \pi)^2]} = 5.075638379964710\text{E-}16$ m
is the radius of a Planck particle

$\xi_m = (\varphi . R_n / a_o)^2 / \Sigma$

$\xi_m = \Sigma . (\rho_u / e)^2$

$\xi_m = \varphi^2 . (4\pi / \xi_v)^4 / \Sigma$

$m_u / m_e . a_o / R_n = \varphi / \Sigma$

If $F^N = G . m_e . m_p / a_o^2$ then; $F^N / F^P = \Sigma$

Its units appear to be m⁶ (volume-squared). I say "appears" because it also unites the Newton and Planck forces (F^N & F^P and φ/Σ; see above), which means it can also have no units.

15.5.2 ρ_u

The ultimate (limiting) density is the maximum possible density that exists in nature, and it applies to both atomic particles; the electron and the proton.

$\rho_u = 7.12660796350449\text{E}+16 \text{ kg/m}^3$

It may be calculated as follows:

$\rho_u = \sqrt{[m_e . m_p]} / V_e^P$
refer to chapter A-6; for V_e^P

or

$\rho_u = a_o . c^2 / G$

or

$\rho_u = \sqrt{[m_e . m_p / \Sigma]}$

15.5.3 G

Today, you will see the units for this constant written as: $N.m^2/kg^2$, which were units of convenience originally assigned to reflect Newton's formula:

$$F = G.m_1.m_2 / R^2$$

This was because its origin, formula and value were unknown, hence its units were unknown.

Planck gave us the means to determine the solution to Newton's gravitational constant 'G' using his formula for 'length' (which is actually Rydberg's orbital radius and EME amplitude 'a_o').

Its Units:

$$\lambda = \sqrt{[G.\hbar / c^3]}$$
$$G = \lambda^2.c^3 / \hbar \qquad \{m^3 / s^2.kg\}$$

Its Value:

$$\lambda = \sqrt{[G.\hbar / c^3]}$$

where: $\hbar = h / 2\pi$ & $\lambda^P = a_o = 5.29177210670E\text{-}11\ m$

$$G = \lambda^2.c^3 / \hbar = 6.67359232004334E\text{-}11$$

Its formula:

$$V_e = 4\pi.\lambda^{P2} / R_n.a_o.\xi_v \ (.\ 1\ m^3)$$
$$\hbar = \tfrac{1}{2}.R_n.m_e.c.\xi_v / 2\pi = R_n.m_e.c.\xi_v / 4\pi$$
$$1/\hbar = 4\pi / R_n.m_e.c.\xi_v = V_e . a_o / 1.m_e.c.\lambda^{P2}$$

Given that; $\lambda^P = \sqrt{[G.\hbar / c^3]}$

$$G = c^3 . \lambda^{P2} . 1/\hbar$$
$$G = c^3 . \lambda^{P2} . V_e . a_o / 1.m_e.c.r^{P2} = c^3/c . a_o . \lambda^{P2}/\lambda^{P2} . [V_e / 1.m_e]$$
$$G = c^2 . a_o [V_e / 1.m_e] \ \& \ [\rho_u = V_e / m_e] \ \& \ [m_u = 1m^3.\rho_u]$$
$$\mathbf{G = c^2.a_o / m_u}$$

confirming that G is a constant, not a factor.

Newton's and Kepler's formulas for the constant of proportionality (K) together gave us the correct units for Newton's gravitational constant (G) 300-years ago;

$$K = (2\pi)^2 / G.m = t^2/a^3$$
$$G = (2\pi)^2.a^3 / t^2.m \quad \{m^3 / s^2.kg\}$$

The Mathematical Laws of Natural Science

Today we have accurate values for Earth's; orbital axes 'a', orbital period (t), and the mass of its force-centre (the sun), so we also have an accurate value for 'G';

$$(2\pi)^2 \times 1.49594598077542\text{E}+11^3 \div (1.9885\text{E}+30 \times 31558118.40^2)$$

$$= \mathbf{6.67359232004334\text{E}-11} \text{ m}^3 / \text{s}^2.\text{kg}$$

And by a remarkable coincidence:

$$c^2.a_o / m_u = \mathbf{6.67359232004334\text{E}-11} \text{ m}^3 / \text{s}^2.\text{kg}$$

confirming that gravity is not based upon the mass of matter, it is based upon the particles of which it comprises.

Conclusion

Despite all of the above;

why is it still necessary to fabricate its units ($N.m^2/kg^2$) from Newton's force-law formula?

why is it still necessary to estimate its value?

why has nobody applied his laws to the atom? because his gravitational constant applies to it.

15.5.4 φ

The coupling ratio is the ratio of magnetic charge energy (E_m) to electrical charge energy (E_e) between a lone proton and a lone electron. It is defined as follows:

$$E_m = G \cdot m_1 \cdot m_2 / R \quad \{J\}$$

$$E_e = k \cdot q_1 \cdot q_2 / R \quad \{J\}$$

$$\varphi = E_m / E_e = G \cdot m_1 \cdot m_2 / k \cdot q_1 \cdot q_2$$

$$\varphi = \frac{(6.67359232004334\text{E-}11 \times 9.1093897\text{E-}31 \times 1.67262163783\text{E-}27)}{(8.98755184732667\text{E+}09 \times 1.60217648753\text{E-}19 \times 1.60217648753\text{E-}19)}$$

$$= 4.40742111792334\text{E-}40$$

If magnetism (*gravity*) causes this coupling ratio to be compromised at the core of a body of matter, its neutrons will revert to their base-particles: alpha (proton) & beta (electron). The kinetic energy (in the alpha and beta particles) released by the neutrons will initiate a chain reaction in all neighbouring neutrons breaking the matter apart. This event can only occur naturally in the ultimate-body, where there is sufficient particle *mass* (magnetic charge) to generate the necessary internal pressure and where the atoms are cold.

The following relationships are also true:

$$\varphi = V_p \cdot a_o / R_n$$

refer to chapter A-6; Table A 6.2 for a definition of V_p

The coupling ratio is the reason we know that matter cannot accrete from hydrogen molecules; the electrical repulsion between proton-electron pairs (hydrogen atoms) is $2.26890050495373\text{E+}39$ ($1/\varphi$) greater than their gravitational (magnetic) attraction. So, our solar system did not, and in fact could not, have accreted from hydrogen - or any other - gas.

$$\varphi = \gamma / \epsilon_p$$

$$\epsilon_p = m_u / m_p$$

15.5.5   k, k', μ_o, ε_o

Coulomb's constant is equivalent to Isaac Newton's *gravitational* constant (G) when applied to *electrical* force in exactly the same fashion. I.e.:

Newton's formula for potential force: $F = G.m_1.m_2 / R^2$

Coulomb's formula for electrical force: $F = k.(q/R)^2$

Their quotient (in a proton-electron pair) is the coupling ratio:

$$\varphi = G.m_p.m_e / k.e^2 = 4.40742111792335\text{E-}40$$

Given the following for the magnetic constant, if:

$\mu_o = 1.0\text{E-}7$ H/m \{kg.m/C²\}

then;

$\mu_o = R_n.m_e / e^2$ \{kg.m/C²\}

and the unit 'Henry' is actually; kg.m²/C²

... and permittivity in a vacuum:

$\varepsilon_o = 1 / \mu_o.c^2 = e^2 / R_n.m_e.c^2$ \{C².s² / kg.m³ = C² / J.m\}

... electrical energy moment.

Coulomb's constant (k) is normally calculated as follows:

$$k = 1/\varepsilon_o = \mu_o.c^2 = 8.98755184732667\text{E+}09 \text{ J.m/C}^2$$

but now it may be calculated thus:

$$k = R_n.m_e.(c/e)^2 = 8.98755184732667\text{E+}09 \text{ J.m/C}^2$$

The problem with Coulomb's constant is that it has been allocated units of convenience, which is because; a) we have still not recognised that mass is magnetic charge, and; b) we cannot (today) calculate *electrical* force. But we can rectify this by multiplying Coulomb's constant by the relative charge capacity squared:

$$k' = \mu_o.c^2 . RC^2 = R_n.c^2/m_e = 2.78024810626745\text{E+}32 \text{ m}^3 / \text{kg.s}^2$$

and replacing electrical charge in his formula with magnetic charge:

$$F_e = k'.(m_e/R)^2 \text{ kg.m/s}^2$$

which not only gives the correct result for electrical force, but also the correct calculation method for the coupling ratio:

The Mathematical Laws of Natural Science

$$F_m:F_e = G.\xi_m.m_e^2 / k'.m_e^2 = G.\xi_m/k' = \varphi = 4.40742111792334\text{E-}40$$

But we still need to allocate more apposite values for magnetic and electrical charges; e.g.

$e = 1$ C & $m_e = 5.68563436731213\text{E-}12$ kg

or

$e = 1.75881869180545\text{E+}11$ C & $m_e = 1$ kg

15.5.6 h, h'

Because we know that energy cannot be created, it can only be transferred, the electro-magnetic energy emitted by a proton-electron pair must be the same as the kinetic energy in the orbiting electron. Moreover, the frequency (f) and amplitude (A) of electro-magnetic radiation is also equal to that of the orbiting electron (A = R and f = v / 2πR).

Max Planck claimed that the energy of electro-magnetic radiation can be calculated using his constant as follows:

$E^P = h.f$

in which his constant (h) is defined thus:

$h = \sqrt{[\pi.m_e.e^2.a_o / \varepsilon_o]} = 6.62607174469163E-34$ **J.s**

However, these units (J.s) can only be correct if a frequency ratio of '1' is applied to his constant thus:

$h = \sqrt{[\pi.m_e.e^2.a_o / \varepsilon_o]} \cdot f_1/f_2$

Therefore, the units for Planck's constant should be:

$\sqrt{[kg.C^2.m / (C^2/m/(kg.m^2/s^2))]} = \sqrt{[kg^2.m^4/s^2]} =$ **kg.m²/s**

$h = \sqrt{[\pi.m_e.e^2.a_o.4\pi.m_e.R_n.c^2 / e^2]}$

$= \sqrt{[4\pi^2.m_e^2.c^2.a_o.R_n]}$

$h = \sqrt{[(4\pi)^2.a_o . R_n]} \cdot \tfrac{1}{2}.m_e.c$ {identical units; kg.m²/s}

Planck therefore *actually* identified a range of orbital radii:

a maximum: $R_o = (4\pi)^2.a_o$ (minimum orbital velocity)

a minimum: R_n (maximum orbital velocity)

and a mean: $R_m = \sqrt{[(4\pi)^2.a_o . R_n]}$

Because we know Planck's mean orbital radius:

$R_m = 4.852618433622630E-12$ m

and we also know the mean velocity: $v_m = \sqrt{[X^R / X.R_m]}$

$v_m = 7224342.80705001$ m/s

we can calculate his minimum electron velocity (v_o) from:

$v_o = c \cdot \sqrt{[R_n / (4\pi)^2.a_o]} = 174090.866621082$ m/s

Given the issue with Planck's units, we can modify his constant thus:
$$h' = h.v_o = \sqrt{(4\pi)^2.a_o.R_n} \cdot \tfrac{1}{2}.m_e.v_o.c$$
Note: $v_o.c = v_m^2$

$$h' = \tfrac{1}{2}.R_n.m_e.c^2 = 1.15353857232684E\text{-}28 \text{ kg.m}^2/\text{s} \cdot \text{m/s (J.m)}$$

Planck's maximum temperature (T_n) is the highest possible in nature. It occurs immediately prior to a proton-electron pair uniting as a neutron.

Planck's minimum temperature (T_o) is that of the gas transition temperature of radon, the largest noble atom.

His mean temperature (T_m) is the mean temperature between T_o and T_n.

Planck:	T_o	T_m	T_n
R (m)	8.35643156381572E-09	4.85261843362268E-12	2.817937953839E-15
v (m/s)	174090.866621084	7224342.80705001	299792459
KE (J)	1.38042005551962E-20	2.37714666443634E-17	4.09355561131261E-14
PE (J)	-2.76084011103925E-20	-4.75429332887267E-17	-8.18711122262522E-14
T (K)	210.193328535837	361962.554671561	623316124.717179

Table 13.5a: Planck's Temperatures

Whilst Max Planck's original formula for electro-magnetic energy ($E^P = h.f$) is incorrect, an alternative approach ($E = h'/\lambda$), using the modified version of his constant (h'), does work.

Another interesting relationship with h': $e^2.k/h' = 2$
$\{C^2 \cdot \cancel{\text{kg.m}^3} / \cancel{C^2.\text{s}^2} \cdot \text{s}^2 / \cancel{\text{kg.m}^3}$ = no units$\}$

Moreover; if we extract $\sqrt{(4\pi)^2.a_o.R_n}$ from h and modify it slightly, thus:
$$\sqrt{(4\pi)^2.a_o / R_n} = 1722.0458764934$$
we get the dynamic ratio ξ_v

It is also interesting to note that the fine-structure constant is equal to:
$$\alpha = e^2 / 4\pi = h'.\varepsilon_o / 2\pi = 2.0427294212227E\text{-}39 \ \{C^2\}$$

Planck's modified constant is related to Coulomb's constant thus:
$$h' = h.v_o = \tfrac{1}{2}.e^2/\varepsilon_o = \tfrac{1}{2}.k.e^2$$
$$= 1.15353857232684E\text{-}28 \ \{\text{kg.m}^3/\text{s}^2 = \text{J.m}\}$$

The Mathematical Laws of Natural Science

If we apply Planck's constants; original and modified to the atom ...

T	R (A)	v	KE	E' (error)	E (error)
6.2332E+08	2.8179E-15	299792459	*4.0936E-14*	1.12193E-11 (274.1)	4.09356E-14 (1)
3.1166E+08	5.6359E-15	211985280.7	*2.0468E-14*	3.96662E-12 (193.8)	2.04678E-14 (1)
2.0777E+08	8.4538E-15	173085256.9	*1.3645E-14*	2.15915E-12 (158.2)	1.36452E-14 (1)
1.5583E+08	1.1272E-14	149896229.5	*1.0234E-14*	1.40241E-12 (137.0)	1.02339E-14 (1)
1.2466E+08	1.4090E-14	134071263.5	*8.1871E-15*	1.00348E-12 (122.6)	8.18711E-15 (1)
1.0389E+08	1.6908E-14	122389758.9	*6.8226E-15*	7.63376E-13 (111.9)	6.82259E-15 (1)
8.9045E+07	1.9726E-14	113310898.8	*5.8479E-15*	6.05785E-13 (103.6)	5.84794E-15 (1)
7.7915E+07	2.2544E-14	105992640.4	*5.1170E-15*	4.95828E-13 (96.90)	5.11694E-15 (1)
6.9257E+07	2.5361E-14	99930819.7	*4.5484E-15*	4.15529E-13 (91.36)	4.54840E-15 (1)
6.2332E+07	2.8179E-14	94802699.6	*4.0936E-15*	3.54785E-13 (86.67)	4.09356E-15 (1)
5.6665E+07	3.0997E-14	90390827.4	*3.7214E-15*	3.07522E-13 (82.64)	3.72141E-15 (1)
5.1943E+07	3.3815E-14	86542628.5	*3.4113E-15*	2.69894E-13 (79.12)	3.41130E-15 (1)
4.7947E+07	3.6633E-14	83147467.9	*3.1489E-15*	2.39359E-13 (76.01)	3.14889E-15 (1)
4.4523E+07	3.9451E-14	80122904.9	*2.9240E-15*	2.14177E-13 (73.25)	2.92397E-15 (1)
4.1554E+07	4.2269E-14	77406080.1	*2.7290E-15*	1.93121E-13 (70.77)	2.72904E-15 (1)
3.8957E+07	4.5087E-14	74948114.7	*2.5585E-15*	1.75302E-13 (68.52)	2.55847E-15 (1)
3.6666E+07	4.7905E-14	72710351.4	*2.4080E-15*	1.60064E-13 (66.47)	2.40797E-15 (1)
3.4629E+07	5.0723E-14	70661760.2	*2.2742E-15*	1.46912E-13 (64.60)	2.27420E-15 (1)
3.2806E+07	5.3541E-14	68777107	*2.1545E-15*	1.35468E-13 (62.88)	2.15450E-15 (1)
3.1166E+07	5.6359E-14	67035631.7	*2.0468E-15*	1.25436E-13 (61.28)	2.04678E-15 (1)
2.9682E+07	5.9177E-14	65420077.9	*1.9493E-15*	1.16583E-13 (59.81)	1.94931E-15 (1)
2.8333E+07	6.1995E-14	63915967	*1.8607E-15*	1.08726E-13 (58.43)	1.86071E-15 (1)
2.7101E+07	6.4813E-14	62511048.9	*1.7798E-15*	1.01712E-13 (57.15)	1.77981E-15 (1)
2.5972E+07	6.7631E-14	61194879.4	*1.7057E-15*	9.54220E-14 (55.94)	1.70565E-15 (1)
2.4933E+07	7.0448E-14	59958491.8	*1.6374E-15*	8.97544E-14 (54.81)	1.63742E-15 (1)
2.3974E+07	7.3266E-14	58794138.4	*1.5744E-15*	8.46263E-14 (53.75)	1.57444E-15 (1)
2.3086E+07	7.6084E-14	57695085.6	*1.5161E-15*	7.99687E-14 (52.75)	1.51613E-15 (1)
2.2261E+07	7.8902E-14	56655449.4	*1.4620E-15*	7.57231E-14 (51.79)	1.46198E-15 (1)
2.1494E+07	8.1720E-14	55670062.1	*1.4116E-15*	7.18404E-14 (50.89)	1.41157E-15 (1)
2.0777E+07	8.4538E-14	54734364.1	*1.3645E-15*	6.82785E-14 (50.04)	1.36452E-15 (1)
2.0107E+07	8.7356E-14	53844315.1	*1.3205E-15*	6.50014E-14 (49.22)	1.32050E-15 (1)
1.9479E+07	9.0174E-14	52996320.2	*1.2792E-15*	6.19784E-14 (48.45)	1.27924E-15 (1)
1.8888E+07	9.2992E-14	52187168.5	*1.2405E-15*	5.91827E-14 (47.71)	1.24047E-15 (1)
1.8333E+07	9.5810E-14	51413982.6	*1.2040E-15*	5.65910E-14 (47.00)	1.20399E-15 (1)
1.7809E+07	9.8628E-14	50674174.4	*1.1696E-15*	5.41831E-14 (46.33)	1.16959E-15 (1)
1.7314E+07	1.0145E-13	49965409.8	*1.1371E-15*	5.19412E-14 (45.68)	1.13710E-15 (1)
1.6846E+07	1.0426E-13	49285576.7	*1.1064E-15*	4.98498E-14 (45.06)	1.10637E-15 (1)
1.6403E+07	1.0708E-13	48632758.7	*1.0773E-15*	4.78950E-14 (44.46)	1.07725E-15 (1)
1.5982E+07	1.0990E-13	48005213	*1.0496E-15*	4.60647E-14 (43.89)	1.04963E-15 (1)
1.5583E+07	1.1272E-13	47401349.8	*1.0234E-15*	4.43482E-14 (43.33)	1.02339E-15 (1)
1.5203E+07	1.1554E-13	46819716.1	*9.9843E-16*	4.27356E-14 (42.80)	9.98428E-16 (1)
1.4841E+07	1.1835E-13	46258980.7	*9.7466E-16*	4.12185E-14 (42.29)	9.74656E-16 (1)
1.4496E+07	1.2117E-13	45717921.4	*9.5199E-16*	3.97810E-14 (41.80)	9.51990E-16 (1)
1.4166E+07	1.2399E-13	45195413.7	*9.3035E-16*	3.84403E-14 (41.32)	9.30354E-16 (1)
1.3851E+07	1.2681E-13	44690421.2	*9.0968E-16*	3.71661E-14 (40.86)	9.09679E-16 (1)
1.3550E+07	1.2963E-13	44201986.6	*8.8990E-16*	3.59608E-14 (40.41)	8.89903E-16 (1)

Table 13.5b: Planck's Energy (E' = h,f)
Note: the above (error) is a ratio and therefore a value of 1 represents zero error

... an error occurs when using Planck's constant 'h'

Whereas there is no error when using the modified version of Planck's constant 'h''

The Mathematical Laws of Natural Science

Max Planck also gave us the crucial link to resolve Isaac Newton's *gravitational* constant (G):

These were Planck's constants; time, length and *mass*:

$t = \sqrt{[\hbar.G/c^5]} = 5.39096122598358\text{E-}44$ s

$\lambda = \sqrt{[\hbar.G/c^3]} = 1.61616952231127\text{E-}35$ m

$m = \sqrt{[\hbar.c/G]} = 2.1765500017459\text{E-}08$ kg

from which, it was possible to complete Planck's atom by defining the associated atomic energy and force formulas; E^P & F^P:

$E^P = \sqrt{[\hbar.c^5/G]}$ {J}

$F^P = c^4/G$ {N}

But Planck did not realise that he had part-defined a fictitious proton-electron pair which led to the solution for 'G':

$G = \lambda^2.c^3/\hbar = \mathbf{6.67359232004333\text{E-}11}$ m³ / kg.s²

from which we can construct the formula for 'G' using Planck's units:

$G = a_o.c^2/m_u = \mathbf{6.67359232004333\text{E-}11}$ m³ / kg.s²

m_u is the unit mass of a Newton atomic particle

Plank's electron orbital properties are defined thus:

	Orbital Radius	Orbital Velocity	Temperature
Maximum R	$R_o = R_n.\xi_v^2$	$v_o = c.\sqrt{[R_n/R_o]}$	$T_o = X.v_o^2$
Mean R	$R_m = R_n.\xi_v$	$v_m = \sqrt{[v_o/c]}$	$T_m = X.v_m^2$
Minimum R	R_n	c	$T_n = X.c^2$
Table 13.5c: The Planck Atom			

The resultant temperatures are:

Radon Gas Transition: T_o = 210.19332853584 K

Mean: T_m = 361962.55467156 K

Neutronic: T_n = 623316124.71718 K

So, whilst Max Planck's energy constant was erroneous, without his work and proposal(s), my solutions would have been much more difficult to achieve; thank you Max! In fact, Max Planck is the *only* 20th Century scientist without whom, none of these scientific discoveries would have been possible.

The Mathematical Laws of Natural Science

15.5.7 e, e'

The electrical charge in an electron is 1.60217648753E-19 C and designated the symbol 'e' (elementary charge unit). This charge is invariable in an electron, but not in a proton:

The static charge in a lone proton is always the same as that in an electron (e), but when a proton attracts an orbiting electron, its charge will increase (e'). This is facilitated by the proton's surplus non-polar magnetic capacity (*mass*). I.e.:

The electrical charge in a lone proton (and in an electron) is always; e

The maximum electrical charge that can be held by a proton is;

$$e_n = m_p . RC = 2.94183820093364\text{E-16 C}$$

which occurs immediately before a proton-electron pair unites to become a neutron:

The charge generated in a proton by its orbiting electron may be defined as follows:

$$e' = e_n . T/T_n \qquad \{C\}$$

It may be concluded, therefore, that; e is used by the proton to maintain its partnership with its orbiting electron, and e' is used to repel adjacent atoms in matter.

15.5.8 R_γ, R_∞, a_o

What is today regarded as *Bohr's* radius (a_o) is the orbital radius of what Bohr referred to as; an electron's *ground state*. But given that ...

$$a_o = 4\pi.\varepsilon_o.\hbar^2 / m_e.e^2 = (h / 2\pi.m_e.c)^2 / R_n \quad \{m\}$$

which breaks down to:

$$a_o = R_n.(\xi_v/4\pi)^2 = 5.2917721067\text{E-}11 \text{ m}$$

... and that a 'ground state' can reasonably be said to occur at the point an electron ceases to provide sufficient kinetic energy to maintain its orbit, this will be when the electron is orbiting at its *maximum* possible orbital radius.

But, given that Planck's minimum and the cold temperature both give us greater orbital radii

$$R = 8.3564315638157\text{E-}09\text{m} \ \& \ 8.59854098572228\text{E-}07\text{m respectively}$$

it is reasonable to assume that 'a_o' does not represent a *ground state*.

Rydberg generated the following formula for his first constant:

$$R_\infty = m_e.e^4 / 8.\varepsilon_o^2.h^3.c = 10973726.9561356 \ \{/m\}$$

which breaks down to:

$$R_\infty = \sqrt{R_n} / 4\pi.a_o^{1.5} \ \{/m\}$$

which breaks down to:

$$R_\infty = 1 / a_o.\xi_v = 10973726.9561356 \ /m$$

and is his wave number (for electro-magnetic energy).

Rydberg generated the following formula for his universal energy constant for an electron:

$$R_\gamma = R_\infty.h.c.(Z.n)^2 = 2.17987197684933\text{E-}18 \text{ J}$$

which breaks down to:

$$R_\gamma = R_n/a_o \cdot \tfrac{1}{2}.m_e.c^2 = 2.17987197684933\text{E-}18 \text{ J}$$

and it occurs when the orbital radius of an electron is equal to; 'a_o'.

Therefore, it was Johannes Rydberg that defined 'a_o' (not Bohr), and this constant should be referred to as *Rydberg's radius*.

The Mathematical Laws of Natural Science

But Planck's maximum orbital radius is; $R_o = [4\pi]^2 \cdot a_o$ {m}

And if Rydberg had modified his formula to reflect Planck's value:

$R_\gamma = \frac{1}{2} \cdot (R_n / a_o \cdot [4\pi]^2) \cdot m_e \cdot c^2 = 1.38042005551962\text{E-}20$ J

Note: $R_\gamma = R_n/a_o \cdot \frac{1}{2} m_e \cdot c^2 = 2.179871976849360\text{E-}18$ J

Rydberg would have revealed Planck's minimum electron kinetic energy 40 years earlier.

The relevance of Rydberg's radius (a_o) to the atomic structure is not yet understood, because electrons have no 'rest-mass'; they leave their proton partners at temperature T_x and velocity 17162.242521927 m/s, and the temperature of the proton-electron pair at a_o is 33192.4000063507 K

15.5.9 R_s

Schwarzschild's radius is the radius of a spherical body, the non-polar magnetic (potential) energy of which, is sufficient at its surface to trap an electron travelling at light-speed (*photon*).

It is calculated as follows:

$R_s = 2.G.m/c^2$
where 'm' is the *mass* of the spherical body.

The following examples give some idea of the Schwarzschild radii of various bodies:

The minimum sized black-body of iron density (ρ_i):

$R_s = c.\sqrt{[\,3\,/\,8.\pi.G.\rho_i\,]} = 1.428750134556622E{+}11$ m

$m = {}^4/_3\pi R_s^3 \cdot \rho_i = 9.6237854E{+}37$ kg

Body of matter of ultimate density (ρ_u):

$R_s = c.\sqrt{[\,3\,/\,8.\pi.G.\rho_u\,]} = 47494.1512680647$ m

$m = {}^4/_3\pi.r^3.\rho_u = 3.19809876372352E{+}31$ kg

Proving that a proton, which has a much smaller mass than 'm' above, cannot trap a *photon* through *gravitational* force (magnetic charge) alone.

A proton:

$R_s = 2.G.m/c^2 = 2.48396784934951E{-}54$ m

Also proving that a proton cannot trap an electron through potential force alone. But if we apply the coupling ratio to Newton's force formula, Schwarzschild's radius is;

$R_s = 2.G.m\,/\,\varphi.c^2 = 5.635587590767792E{-}15$ m

i.e. exactly twice the neutronic radius ($R_n = 2.81793795383896E{-}15$ m)

However:

The Schwarzschild radius (R_s) is fictitious because an electron in free-flight cannot travel at velocity 'c' and there are no such things as *photons*.

15.5.10 R_n, t_n

The neutronic radius is the orbital radius of an electron when it is travelling at velocity; 'c'. At this speed, the electron and proton combine to become a neutron.

$R_n = \mu_o.e.RC$ {kg.m/C² . C . C/kg = m}

$R_n = G.m_p / \varphi.c^2$ {m³ / kg.s² / kg . s²/m² = m}

$R_n = \mu_o.e.RC = G.m_p / \varphi.c^2 = 2.81793795383896\text{E-}15$ m

The *gravitational* acceleration in the proton-electron pair according to Newton and Coulomb at this time:

$g_n = c^2/R = 3.18940728807838\text{E+}31$ m/s²

$g_n = G.m_p / \varphi.R_n^2 = 3.18940728807838\text{E+}31$ m/s²

The time an electron takes to orbit its proton at '*light-speed*' is;

$t_n = 2\pi.R_n/c = 5.90596121302193\text{E-}23$ s

15.5.11 RAC, RAM

Relative atomic charge and relative atomic *mass* define the capacity of a particle to hold an electrical charge (e) or magnetic charge (*mass* 'm') respectively.

They are related as follows:

   RAM {g/mol}

   RAC {C/mol}

Divide either of the above by Avogadro's constant (N_A), you will get the capacity (m, e) of the particle {g, C}

Divide the ideal gas constant (R_i) by either of the above and you will get the *specific* capacity (c, q) of the particle {J/g/K, J/C/K}

Multiply the capacity by the *specific* capacity you will get the *relative* capacity (C, Q) of the particle {J/K} – Boltzmann's constant

The Mathematical Laws of Natural Science

15.5.12 N_A

Avogadro's number is the number of C^{12} (pure carbon) atoms in 12g and is recognised as; $N_A = 6.02214129E+23$ /mole

However, one atom of pure carbon-12 has a *mass* of:

$m_C = 6.(m_e + m_p + m_n) = 2.00823909216E-23$ g

i.e.; $N_A = 1/(m_e + m_p)$ {/mol}

where m_e & m_p are specified in grams

and 12 grams of pure carbon-12 contains:

$N_A = 12/m_C = 5.97538412973187E+23$

which is 0.7764208% less than Avogadro's number

If corrected, this would, of course alter a number of constants such as:

R_i; X; X^R; c & q

However, throughout this book, all property values have been based upon Avogadro's value; *Avogadro's Number* has been left as *he* defined it.

The Mole:

A mole of matter (e.g. Quanta, atoms, molecules, etc.) is the mass that contains Avogadro's number (N_A) of particles of that matter. It also happens to be the RAM of that particle in grams.

$RAM = m_a . N_A$ {g/mole}

Convert mass to moles:

$n = m/RAM$ {g ÷ (g/mole) = moles}

Convert moles to mass:

$m = n.RAM$ {moles x (g/mole) = g}

where:
RAM = relative atomic mass
m_a = atom or molecule mass
NA = Avogadro's number
n = number of moles
m = mass

E.g. a mole of water (H_2O) is 18.01528g (2x1.00794 + 15.9994)

The Mathematical Laws of Natural Science

15.5.13 k_B, R_i

Boltzmann's constant (k_B) defines the kelvin temperature scale based upon the potential energy of a proton-electron pair. He defined this scale thus:

$$k_B = 1.38065156\text{E-}23 \text{ J/K}$$

which means that at the neutronic condition, the temperature of the proton-electron pair is:

$$\underline{T}_n = m_e.c^2 / Y.k_B = 6.23316124717170\text{E+}08 \text{ K}$$

If we wish to alter the temperature-scale we simply need to select our preferred neutronic temperature; say 1.0E+08 K, and recalculate his constant like this:

$$k_B = m_e.c^2 / Y.\underline{T}_n = 8.60582379963926\text{E-}23 \text{ J/K}$$

Together with Avogadro's constant, Boltzmann's constant defines the ideal gas constant:

$$R_i = N_A.k_B = 8.249923420031355 \text{ J/K/mol}$$

The Mathematical Laws of Natural Science

15.5.14 K & h

Newton's constant of proportionality is common to all orbits and universal expansion.

In elliptical orbits, its value is governed by the force-centre's *mass*;

$$K = t^2/a^3 = (2\pi)^2 / G.m_1 = (2\pi)^2 / (v_{max}.v_{min}.a) \quad s^2/m^3$$

where:
v_{max} = maximum orbital velocity (@ orbital perigee)
v_{min} = minimum orbital velocity (@ orbital apogee)
a = half the length of the orbital major axis

In circular orbits its formula looks like this:

$$K = (2\pi/v)^2 / R \qquad \{s^2/m^3\}$$

Where:
v = orbital velocity
R = orbital radius

In the post 'Big-Bang' expansion of the universe, its value is also governed by the force-centre's *mass*;

$$K = (2\pi)^2 / G.m_1 = (2\pi)^2 / 2.R_o.(u/2)^2 = 2.\pi^2 . t_o^2/R_o^3 \quad \{s^2/m^3\}$$

Where:
R_o = the ultimate distance travelled by all universal matter.
t_o = the time taken for all universal matter to cease outward expansion.
u = the initial velocity of all universal matter; 1,773,498.41 {m/s}

Its value {s²/m³} can be any number greater than 0 and less than 1

For a proton-electron pair: K = 0.15587874533403
For our lunar system: K = 9.91826542816423E-14
For our solar system: K = 2.97491436434708E-19
For our Milky Way: K = 3.350257445662553E-30

Assuming an orbital eccentricity of 0.015941744 for our solar system

For the universe: K = 5.669761532297553E-35

Assuming a current universal age of 13.77 billion years.

Newton's constant of motion of an elliptical orbit is calculated like this;

$$h = R.v \qquad \{m^2/s\}$$

and for universal expansion, it is calculated like this;

$$h = R.u/2 \qquad \{m^2/s\}$$

15.5.15 ξ_v, ξ_m

The **static ratio** defines the relationship between the *mass* of a proton and the *mass* of an electron, but because both have the same density, it also describes their relative volumes:

$$\xi_m = m_p/m_e = V_p/V_e = 1836.15115053207 \qquad \text{no units}$$

This value is very specific. It defines the orbital instant when magnetic field energy generated in the proton-electron pair equals the centrifugal energy in the electron. Which occurs when the electron is orbiting at the velocity of electro-magnetic energy (c) and simultaneously achieving the neutronic radius (R_n).

If ξ_m > the above value; the orbiting electron would impact its proton partner before it achieved 'c'

If ξ_m < the above value; the orbiting electron's velocity would achieve 'c' before it achieved R_n and thereafter remain constant

In both cases, neutrons could not be created and our universe would not exist.

The **dynamic ratio** defines the relationship between '*light-speed*' velocity (c) and Planck's minimum velocity (v_o):

$$\xi_v = c/v_o = 1722.0458764934 \qquad \text{no units}$$

It was found from:
Planck's constant; $h = \sqrt{(\pi.m_e.e^2.a_o / \varepsilon_o)} = 6.62607174469163E\text{-}34 \text{ kg.m}^2/\text{s}$
and;
Rydberg's constant; $a_o = \varepsilon_o.(h/e)^2 / \pi.m_e = 5.2917721067E\text{-}11 \text{ m}$
But is derived from:

$$\xi_v = 4\pi.\sqrt{[a_o/R_n]} = 1722.0458764934$$

The Mathematical Laws of Natural Science

It has since been discovered that every property of the proton-electron pair is in some way related to the dynamic ratio.

	Minimum (m_1):	(m_2) Mean (m_1):	(m_2) Maximum:	units	formula:	M:M
T	210.19332853584	361962.554671561	623316124.717178	K	m_2/m_1	ξ_ν
R	8.35643156381571E-09	4.85261843362263E-12	2.81793795383896E-15	m	m_1/m_2	ξ_ν
v	174090.866621084	7224342.80705004	299792459	m/s	$\sqrt{[m_2/m_1]}$	ξ_ν
G	3.62686268767106E+18	1.07552509449653E+25	3.18940728807838E+31	m/s²	$\sqrt{[m_2/m_1]}$	ξ_ν
T	3.01595419916531E-13	4.22043937529269E-18	5.90596121302193E-23	s	$^2\!/\!_3\sqrt{[m_1/m_2]}$	ξ_ν
KE	1.38042005551962E-20	2.37714666443636E-17	4.09355561131267E-14	J	m_2/m_1	ξ_ν
PE	-2.76084011103925E-20	-4.75429332887272E-17	-8.18711122262534E-14	J	m_2/m_1	ξ_ν
F	-3.30385056103851E-12	-9.79737721789819E-06	-2.90535538991261E+01	N	$\sqrt{[m_2/m_1]}$	ξ_ν
H	1.45477841280E-03	3.50569790763E-05	8.44796548491E-07	m²/s	$[m_1/m_2]^2$	ξ_ν
F	3.31570021944218E+12	2.36942154851033E+17	1.69320448260839E+22	Hz	$^2\!/\!_3\sqrt{[m_2/m_1]}$	ξ_ν
λ	9.04160325599145E-05	1.26525589837942E-09	1.77056263481047E-14	m	$^2\!/\!_3\sqrt{[m_1/m_2]}$	ξ_ν
V	0.17231810182	296.73967667583	510999.3366116	J/C	m_2/m_1	ξ_ν
I	2712.81241061556	2712.81241061556	2712.81241061556	C/s	m=m	1
R	6.35200949172824E-05	1.09384517526775E-01	1.88365157359204E+02	J.s/C²	m_2/m_1	ξ_ν
Proton-Electron Pair Properties and the Static Ratio						
M:M refers to both 'mean:minimum' and 'maximum:mean' ratios						

15.5.16 B, RC

The magnetic field 'B' as described by Lorentz is actually 1/RC, where RC is the relative charge capacity of Quanta. And the orbital radius at which an electron and a proton combine to create a neutron may be calculated using it:

$$R_n = \mu \cdot e / B \quad \{m\}$$

Lorentz's magnetic field constant was originally derived as follows:

$$B = \mu_o \cdot I / 2\pi \cdot R \quad \{kg/C\}$$
where: $R = 2 \cdot R_n$ & $I = e$

But can also be calculated thus:

$$B = 1/RC = m_e/e = 5.685634367312\text{E-}12 \text{ kg/C}$$

RC is the relative [electrical] charge capacity of matter of ultimate density (i.e. Quanta), but may also be applied to any *mass*; you simply need to rationalise its density:

Relative charge capacity of planet earth; $RC_E = e/m_e \cdot \rho_E/\rho_u \quad \{C/kg\}$

The Mathematical Laws of Natural Science

15.5.17 X, X^R

X & X^R are both [heat] energy transfer coefficients for the electron.

$\underline{T} = PE / k_B \cdot \sqrt[3]{[\, ½ \cdot \sqrt{[(4\pi)^2 \cdot a_o / R_n]}\,]}$ {K}
where $\sqrt[3]{[\, ½ \cdot \sqrt{[(4\pi)^2 \cdot a_o / R_n]}\,]} = Y$

$\underline{T} = PE / Y \cdot k_B$ {K}

By applying, Gilbert's, Newton's and Coulomb's force formulas:

$\underline{T} = X_e \cdot v^2 / e^2$ {K.C^2.(s/m)^2 . (m/s)^2 / C^2}

$X_e = \underline{T} \cdot e^2 / v^2$ {K.C^2 . (s/m)^2}

To find the value of X; the neutronic condition for X_e:

$X_e = \underline{T}_n \cdot (e_n / \xi_m)^2 / c^2$ {K . (C.s/m)^2}

Note: $e_n = m_p \cdot RC$ but this calculation is for the velocity (and electrical charge) of an electron: hence; e_n/ξ_m

because 'e' is a constant:

$X = \underline{T}_n \cdot (e_n / \xi_m / e \cdot c)^2 = \underline{T}_n / c^2 = 6.93532716478932E\text{-}09$ {K . (s/m)^2}

therefore:

$\underline{T} = X \cdot v^2$ {K}

X {K.s^2/m^2} is an energy transfer coefficient for an electron. As an orbiting electron receives electro-magnetic energy, it converts this energy into velocity according to the relationship:

$v = \sqrt{[\underline{T}/X]}$ {m/s}

X^R {K.m} is also an energy transfer coefficient for an electron. This coefficient is another way of writing X but converting electro-magnetic energy into orbital radius:

$R = X^R / \underline{T}$ {m}

Moreover, if:

$\underline{T} = X \cdot v^2 = X^R / R$ {K}

$R = X^R / X \cdot v^2$ {m}

At light-speed (c)

$X^R / X \cdot c^2 = R_n = 2.81793795383896E\text{-}15 \text{ m}$

According to the relationship: $\underline{T} = X \cdot v^2$

$\underline{T}_n = X \cdot c^2 = 623316124.71718 \text{ K}$

The Mathematical Laws of Natural Science

which is achieved by a proton-electron pair at the instant of their union as a neutron, it therefore represents the highest possible temperature achievable by natural means.

This is the temperature at the centre of all *bright* stars, i.e. those that have collected sufficient satellite matter to generate fissionable energy, making them visible to the naked eye in the night sky.

You may have noticed that;

$\underline{T} = X.v^2 / e^2$ is similar to Newton's *gravitational* force;

$F = G.m_1.m_2 / R^2$, Coulomb's force; $F = k.e^2 / R^2$, and Gilbert's and Maxwell's formulas for force and energy. It is therefore anticipated that all of these formulas will eventually become just two; one for magnetic charge (*mass*) and the other for electrical charge.

An interesting relationship for the above heat constants is as follows:

$4\pi^2 . X/X^R = \mathbf{K}$ $\{s^2/m^3\}$

*Where: **K** is Isaac Newton's orbital constant of proportionality for circular orbits (for the atom):*

$K = t^2/a^3 = 0.15587874533403 \; s^2/m^3$

15.5.18 Y

Question: To what does the heat transfer constant apply in atomic physics? @ 293K ...

The potential energy in a proton-electron pair (PE_{pep}):

$PE_{pep} = m_e.v^2 = 3.940430E\text{-}20$ J

$PE_{pep}/Y = 4.045309E\text{-}21$ J

And the potential energy between adjacent atoms in elemental matter (PE_{em}):

$p_{em} = R_i.T_.\rho/(RAM/1000) = 8.7390993E\text{+}07$ N/m²

$PF_{em} = R_i.T_.\rho/(RAM/1000) . d^2 = 1.117821E\text{-}11$ N

$PF_{em} = R_i.T_.m_a/d^3/(RAM/1000) . d^2 = 1.117821E\text{-}11$ N

$PF_{em} = k_B.N_A.T_.m_a/d^3/(RAMRAM/1000) . d^2 = 1.117821E\text{-}11$ N

$PE_{em} = k_B.N_A.T_.m_a/d^3/(RAM/1000) . d^3 = 4.045309E\text{-}21$ J

$PE_{em} = k_B.N_A.T_.m_a/(RAM/1000) = 4.045309E\text{-}21$ J

$N_A.m_a/(RAM/1000) = 1.0$ (all elements at any temperature)

$PE_{em} = k_B.T_. 1 = 4.045309E\text{-}21$ J

$PE_{em} = k_B.T_ = 4.045309E\text{-}21$ J

Therefore, potential; energy, force and pressure between adjacent atoms in any elemental matter are all less than those in a proton-electron pair (at the same temperature) by the heat transfer constant (Y).

Because $PE_{em} = k_B.T_ = PE_{pep}/Y$, and because $p_{em} = R_i.T_.\rho/(RAM/1000)$ is the universally accepted formula for calculating the pressure in a gas, i.e. the repulsion between adjacent atoms, we know that 'Y' must apply to the electrical force of repulsion (F_e) between adjacent atoms in elemental matter

$F_e . d/R_1 = F_{pep}/Y$

rather than the magnetic attraction (F_m):

$F_m = \mu_o.\xi_m . I_1^2 . (2\pi)^2 . (R_1/d)^3 . RAM.m_n/m_p / \zeta^3 = \rho.h_e^2 / \zeta^3$

The question is, why does the heat transfer constant apply to elemental matter and not to the proton-electron pair?

However, the potential energy in a proton-electron pair is dynamic, whilst the potential energy between adjacent atoms is static, and 'Y' is dependent upon the dynamic ratio: $Y = \sqrt[3]{[½.\xi_v]}$

Appendices

These appendices provide support for the main body of the book.

The Mathematical Laws of Natural Science

A-1 Glossary

Apogee (aphelion)	the point at which a satellite's orbit passes furthest from its force-centre. The term apogee normally applies to lunar orbits, and aphelion is normally used for planetary orbits, but they mean the same thing.
atom	a collection of deuterium and tritium atoms that were fused together in core of massive, cold bodies.
atomic particles	the proton and the electron
charge (electrical)	the electrical potential of an atomic particle
charge (magnetic)	the magnetic potential of an atomic particle
coupling ratio	the ratio of gravitational and electrical charge forces in a proton-electron pair.
crystal	same-element matter the atoms of which are arranged in their lattice structure
current (electrical)	the rate of flow of electrons
deuterium	a hydrogen atom with one neutron attached
electricity	electron flow rate
EME	the electro-magnetic energy generated by a proton-electron pair
energy	the distance over which a force is applied
force	the effort required to change the velocity and/or direction of a mass
force-center	the central body (anchor) of an orbital system.
galactic force-centre	the force-centre of a galaxy.
gaseous (matter)	matter in which its atomic electrical charge repulsion forces are greater than its magnetic field force

The Mathematical Laws of Natural Science

gravity	non-polar magnetic attraction
(the) great attractor	the residue left over from the last 'Big-Bang', that is slowing down the outward travel of all universal galaxies (through magnetic attraction), and will eventually cause them to reaccrete into another ultimate body.
Hades	the name adopted here for the milky way's force-centre
heat	the electro-magnetic energy absorbed by a body's atoms
hydrogen	a proton-electron pair
inter-atomic	between adjacent atoms
kinetic energy	the energy in a mass moving at a constant velocity
lattice structure	the unique arrangement of atomic nucleic protons, that is replicated between adjacent atoms in both viscous and gaseous states
light	the electro-magnetic energy detected by a living body's optical receptors
liquid	viscous matter that cannot hold its shape under the ambient pressure and gravitational force
mass	non-polar magnetic charge
massive body	cold bodies with sufficient mass to generate the core pressure necessary to fuse atoms; e.g. galactic force-centres, the great attractor and the ultimate body.
matter	collection of atoms
neutron	proton-electron pair united through high temperature
neutronic	the condition of a proton-electron pair at the instant of its union as a neutron

The Mathematical Laws of Natural Science

optical receptors	the means whereby a living body can 'see' electron-magnetic energy
orbit	the curvilinear path followed by a satellite about its force-centre
orbital system	a force-centre and all of its satellites.
Perigee (perihelion)	the point at which a satellite's orbit passes closest to its force-centre. The terms perigee normally applies to lunar orbits, and perihelion is normally used for planetary orbits, but they mean the same thing.
potential energy	the straight-line attraction or repulsion between masses
power	the rate at which energy is applied
pressure (attraction)	magnetic field force (per unit area)
pressure (repulsion)	electrical charge force (per unit area)
proton-electron pair	a proton with a single orbiting electron
Quanta	the collective term for atomic particles
resistance (electrical)	the potential force holding an electron in its atomic shell
resistivity (electrical)	the potential energy needed to transfer an electron between atoms
same-element matter	matter that contains 100% atoms of the same atomic number (Z)
satellite	an orbiting body.
state of matter	matter that is either viscous or gaseous
temperature	a mathematical definition of the electro-magnetic energy radiated by a proton-electron pair

The Mathematical Laws of Natural Science

temperature of an atom	the electro-magnetic energy radiated by an atom's proton-electron pairs, the electrons of which, are orbiting in shell-1
tritium	a hydrogen atom with two neutrons attached
(the) ultimate body	all universal matter that has reaccreted at the end of a universal period due to the gravitational attraction of the great attractor.
universal period	the period of time between 'Big-Bangs' (64.75 bn years)
viscous (matter)	matter in which its atomic electrical charge repulsion forces are less than its magnetic field force
Voltage (electrical)	the potential force (difference) required to pull electrons from their atomic shells

A-2 References

There is little in today's scientific literature that has, or can, help to resolve the mathematical laws of natural science. Therefore, apart from the achievements of the heroes listed in Appendix A-4 of this book, most of these laws have been established as a result of the work done in the previous publications listed below:

Philosophiæ Naturalis Principia Mathematica Rev. IV; Keith Dixon-Roche; 978-1-07215-605-5

The Atom; Keith Dixon-Roche; 978-1-08610-029-7

The Neutron; Keith Dixon-Roche; 978-1-08251-683-2

The Physical Constants; Keith Dixon-Roche; 978-1-79422-609-8

The Life & Times of the Neutron; Keith Dixon-Roche; 978-1-08239-479-9

The Universe; Keith Dixon-Roche; 978-1-70753-878-2

Some additional references used in the creation of this book are listed below:

Magnificent Principia; Colin Pask; 978-1-61614-745-7

Seven Brief Lessons on Physics; Carlo Rovelli; 978-0-141-98172-7

Science Data Book; Open University; 0 05 002487 6

Science and Technology Dictionary; Chambers; 0-550-18026-5

A Dictionary of Scientific Units; H G Jerrard & D B McNeill; 0-412-28100-7

It is important to note here that most of the sources used in this work are from work done by pre-20th Century scientists that are universally known and available from sources too numerous to mention here.

The Mathematical Laws of Natural Science

The following publications refer to the documented values for the gas-transition temperature of the atomic elements (Chapter 8.2):

SMR: Smithell's Metals Reference Book; E A Brandes & G A Brook; Butterworth

HME: Handbook of Metal Etchants; Perrin Walker, William H. Tarn; CRC Press

SDB: Science Data Book; R M Tennent; Open University

MEH: Mechanical Engineer's Handbook; Myer Kutz; Wiley

ELE: The Elements; Theodore Gray & Nick Mann; Black Dog & Leventhal Publishers Inc.

MSH: Mark's Standard Handbook for Mechanical Engineers; Theodore Baumeister; McGraw Hill

CEH: Chemical Engineer's Handbook; Robert H Perry & Cecil H Chiltern; McGraw Hill

CRC: Handbook of Chemistry and Physics; W M Haynes; CRC Press

The Mathematical Laws of Natural Science

A-3 Useful Formulas

Equidistant arc-length between 'n' points on the surface of a sphere:
$d = \pi \cdot A / C \cdot n$
where C is the circumference of the sphere
Linear distance across arc-length 'd' (above):
$\ell = 2 \cdot R \cdot \sin(½ \cdot d/R)$
but if you know 'ℓ' and need to find 'n':
$n = \pi / \text{Asin}(½ \cdot \ell/R)$
and if $\ell = R$:
$n = \pi / \text{Asin}(½) = 6$

Lorentz's Equation (magnetic force or field strength):
$F = q \cdot v \cdot B$
Which becomes:
$F = q \cdot g \cdot R \cdot B$
for the laws of orbital motion

> *where:*
> *q is the total electrical charge* $= q_1 \cdot q_2 / m_e \cdot (q_1 + q_2)$
> *v = relative velocity (electrical circuits)*
> *g = gravitational attraction between m_1 & m_2*
> *R = radial separation between m_1 & m_2*
> $B = \mu_o \cdot e/R_n = R_n \cdot m_e/e^2 \cdot e/R_n = m_e/e = 1/RC \text{ kg/C}$
> *RC is the relative atomic charge capacity of an electron {C/kg}*
> $B = 1/RC = 5.685634367312\text{E-}12 \text{ kg/C}$

Inter-atomic force factor (F_T):
$\underline{T}_k = \underline{T}_n / \xi_m \cdot Y^2$
$F_T = \underline{T}_1 / \underline{T}_k$

$\underline{T}_1$ = measured temperature of atom (shell-1 temperature)

A-4 The Heroes

The heroes of this story, to which I offer my gratitude, are listed below

It is not necessary to identify the invaluable contributions made by each of these contributors, they are all widely known and available in almost every scientific publication in circulation today.

Nicolaus Copernicus (Polish) 1473-1543
William Gilbert (English) 1544-1603
Tyco Brahe (Danish) 1546-1601
Galileo Galilei (Italian) 1564-1642
Johannes Kepler (German) 1571-1630
Christiaan Huygens (Dutch) 1629-1695
Isaac Newton (English) 1642-1727
Edmund Halley (English) 1656-1741
Charles-Augustin de Coulomb (French) 1736-1806
Hans Christian Ørsted (Danish) 1777-1851
Michael Faraday (English) 1791-1867
Jospeh Henry (USA) 1797-1878
Josef Stefan (Austria) 1815-1863
James Clerk Maxwell (Scottish) 1831-1879
William Crookes (English) 1832-1919
Ludwig Boltzmann (Austria) 1844-1906
Hendrik Lorentz (Dutch) 1853-1928
Jules Henri Poincaré (French) 1854-1912
Johannes Robert Rydberg (Swedish) 1854-1919
Max Karl Ernst Ludwig Planck (German) 1858-1947

The others that were instrumental in the completion of this book are:

My long-suffering wife (Brigitte) sub-editor and critic

My daughter (Eléonore), who initiated this project

Kenneth Pickering friend & editor, who first suggested that I write it

My thanks go out to all the above each of whom have provided a valuable piece of the puzzle without which the final solution would not have been possible, along with my sincere apologies to anybody I have unintentionally omitted.

A-5 Newton's Orbital Laws

The four principal agents for the theories of planetary motion were Copernicus, Kepler, Galileo and Newton. Between them, they defined the behaviour of orbiting satellites, moons and planets that remain valid even today.

A-5.1 Nicolaus Copernicus (1473 to 1543)

Copernicus stated that; contrary to religious doctrine, the sun does not orbit the earth, but all the planets in the solar system orbit the sun. He was so concerned for his safety regarding this claim, however, that he arranged for the publication of his findings to be deferred until after his death.

A-5.2 Johannes Kepler (1571 to 1630)

Kepler used Tycho Brahe's (1546 to 1601) observational data to show that the planets not only orbited the sun, just as Copernicus had previously claimed, but that their orbital paths were ellipses. Kepler also stated that the time taken to traverse between any two points on this elliptical curve is proportional to the swept area:

i.e; $t_1/A_1 = t_2/A_2$

Whilst he did not provide a mathematical proof for his swept area theory, he understood it. It was later confirmed by Isaac Newton.

A-5.3 Galilei Galileo (1564 to 1642)

Galileo is best known for his physical evidence of celestial bodies (moons) orbiting other planets, revealed in his book; Dialogue Concerning the Two Chief World Systems (frequently referred to as the 'Dialogue'), therein declaring Copernicus correct and finally quashing over a thousand years of religious dogma that stated all celestial bodies orbit the earth. In return for his findings, he was put under permanent house arrest, but only after being threatened with death if he didn't recant this claim.

However, it was during his confinement that Galileo completed his most important work, his laws of motion, one of which states that a body fired from the surface of the earth would follow a parabolic curve back to its surface.

The Mathematical Laws of Natural Science

This claim may be demonstrated by comparing Galileo's mathematically correct parabola with a projectile trajectory calculation:

$x(t) = A.t + B$

$y(t) = C.t + D - \frac{1}{2}.g.t^2$

If B and D are zero {i.e. v occurs at t = 0}:

$x(t) = v.Cos(\alpha).t$

$y(t) = v.Sin(\alpha).t - \frac{1}{2}.g.t^2$

Where:

v = initial velocity

A = initial horizontal velocity {i.e.; $A = v.Cos(\alpha)$}

B = offset horizontal distance from t = 0

C = initial vertical velocity {i.e.; $C = v.Sin(\alpha)$}

D = offset vertical distance from t = 0

This plot shows the projectile trajectory (curve) superimposed on two alternative parabolic curves, one of which passes through the same latus rectum and the other being the best parallel match.

Whilst the parabolic path is not strictly correct, it is stunningly close, demonstrating that given the limited information and facilities available to Galileo at his time, he was a very capable mathematician.

The Mathematical Laws of Natural Science

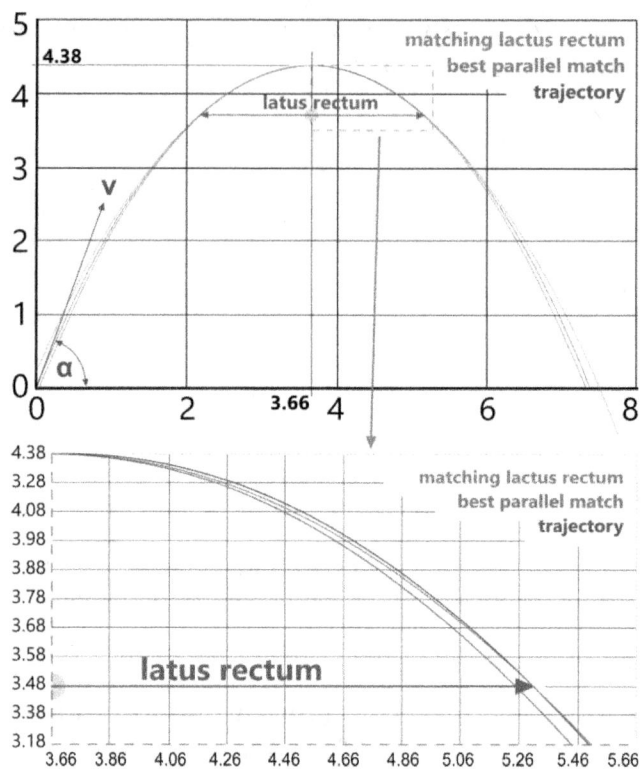

The Mathematical Laws of Natural Science

A-5.4   Isaac Newton (1642 to 1727)

Along with [his explanation for] *gravity*, Newton used [his creation of] calculus to mathematically prove the theories previously generated by Copernicus, Kepler and Galileo. In 1687 he published his results under the heading Philosophiæ Naturalis Principia Mathematica (first of three issues), probably the most important scientific work ever produced.

$$F = \frac{G.m_1.m_2}{r^2}$$

In his Principia, Newton discusses the alternative curves that describe the elliptical paths followed by an orbiting body. However, the parabolic and hyperbolic curves can only be responsible for paths followed by a body (e.g. a galactic comet) travelling towards a force centre from well outside its influence, sufficiently close to fall under its influence, pass around the force centre and then travel back out of its influence. A complete orbit, i.e. that of a satellite must be an ellipse.

As a result of this work, Newton defined the fundamental relationship (G) between attracting bodies in which the potential force (F) is directly proportional to the inverse of the square of the distance (R) between the attracting bodies.

Whilst a value for 'G' was never established by Newton, despite it being of special importance to his theories, it has been estimated many times since the publication of Principia, varying between 6.67E-11 and 6.76E-11 N.m²/kg²

The minimum and maximum radial distances between the earth and sun (@ a & b respectively) are assumed to be as defined in the Earth-Sky fact sheet (https://earthsky.org/). Therefore, using Newton's theories and true value for 'G', the principal properties of the earth's orbit are as follows:

a = 1.495945981E+11m (R + R)/2)
b = 1.495737135E+11 {√[a².(1-e²)]}
e = 0.01670914665 {a.e² + R.e + R - a = 0}
p = 1.495528319E+11m {a.(1-e²)}
f = 1.47095E+11m {a.(1-e)}
x' = 2.499598078E+09m {a-*f*}
R = 1.47095E+11m to 1.520941962E+11m
F = 3.658178805E+22 to 3.421649078E+22N
v = 30286.008788376 to 29290.53557m/s

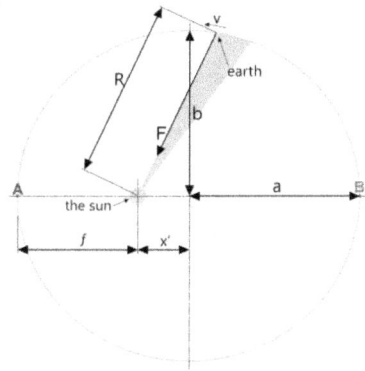

The Mathematical Laws of Natural Science

Newton's creation of Calculus allowed him to generate formulas for non-linear versions of Galileo's relationships for distance (s), time (t), velocity (v) and acceleration (a) as follows:

$s = ut + \frac{1}{2}at^2$

$\delta s/\delta t = v = u + at$

$\delta^2 s/\delta t^2 = \delta v/\delta t = a$

The Mathematical Laws of Natural Science

A-5.5   Proof (elliptical orbits)

By applying calculus, Newton was able to generate the non-linear formulas necessary to complete his theories concerning the elliptical (conic) path of orbiting bodies, which was proven as follows:

Assume an ellipse and the planet is passing the x-axis @ 'A' (y = 0)

x component = R {a}

y component = v/ω {b}

$x(t) = R \cdot \sin(\omega \cdot t)$

$y(t) = (v/\omega) \cdot \cos(\omega \cdot t)$

From: $\sin^2(\omega \cdot t) + \cos^2(\omega \cdot t) = 1$

$y(t)^2 / (v/\omega)^2 = 1 - \sin^2(\omega \cdot t)$

$\sin^2(\omega \cdot t) = 1 - y(t)^2 / (v/\omega)^2$

$x(t)^2 / R^2 = \sin^2(\omega \cdot t) = 1 - y(t)^2 / (v/\omega)^2$

$x(t)^2 / R^2 = 1 - y(t)^2 / (v/\omega)^2$

$x(t)^2 / R^2 + y(t)^2 / (v/\omega)^2 = 1$

An ellipse!

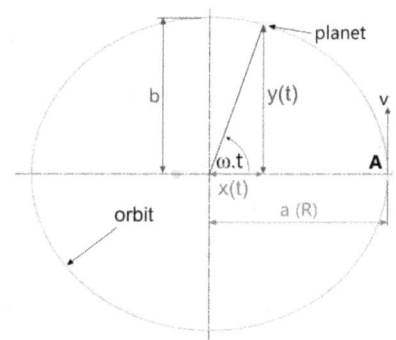

A-5.6 Euclidean Geometry (equal areas)

Whilst Kepler had already predicted the equal-swept-area-with equal-orbital-time theory, it had still not been mathematically proven by the time Newton was writing his Principia. Newton did this using Euclidian geometry.

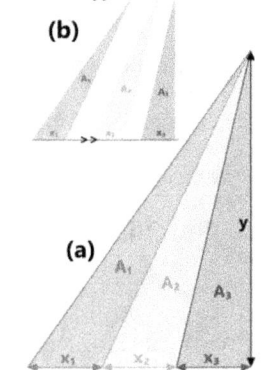

The areas of each triangle; A_1, A_2 & A_3 are all equal if the base widths; x_1, x_2 & x_3 are equal, which can be proven as follows:

Let y=6 and x_1, x_2 & x_3 all equal 3 (a)

The area of a triangle: $A = x.y/2$

$A_1 = 6 \times 3 \div 2 = 9$

$A_2 = 6 \times (3+3) \div 2 - A_1 = 9$

$A_3 = 6 \times (3+3+3) \div 2 - A_1 - A_2 = 9$

Therefore, all the areas are equal (i.e. 9)

The same applies to triangles with equal bases between parallel lines (b)

He then applied this to the conservation of energy

The Mathematical Laws of Natural Science

A-5.6.1 Proof (conservation of energy & equal time-swept area)

Newton's proposition diagram for his proof of Kepler's 'equal-areas-equal-time' theory is shown here, where the following instructions describe its construction (author's words):

1) Divide time [of orbit] into equal parts [represented by equal swept areas {triangles}]

2) Assume the line A-B describes the linear path of the body if unconstrained by potential attraction

3) The same body would then continue to B-c

4) Assume that the body is attracted by a central-force (S) and diverted from its right line (B-c) in a direction parallel to V-B as far a C

5) Continue to generate similar triangles (S-A-B) following the points D, E, F, etc.

Note: The dimensions L, θ, X_o & Y_o in the above diagram were not part of Newton's original drawing. They have been added by me in order to assist with the correlation between all three figures in this chapter.

Newton was therefore stating that all swept areas (triangles SAB, SBC, SCD, SDE, SEF, etc.) must be equal.

The difficulty in generating the above diagram is knowing how far along the line c-C that C occurs in order to ensure that each subsequent area remains equal.

This can be achieved using the process described in the figure below, where the blue variables are entered (X_o, Y_o, L, θ), all the green variables (X_1, Y_1, R_1, r_2, A_2, ε_2, X_o, α_1) can be easily calculated using the blue variables, and the red variables (h, R_2, α_2, X_2, Y_2) may be determined using the formulas provided.

$h = r_2 \cdot Tan(\varepsilon_2)$
$R_2 = 2 \cdot A/h$
$\alpha_2 = [h/R_2 \cdot Sin(½\pi - \varepsilon_2)]$
$X_2 = R_2 \cdot Cos(\Sigma\alpha)$
$Y_2 = R_2 \cdot Sin(\Sigma\alpha)$

Page 237

The Mathematical Laws of Natural Science

Newton claimed that if you reduce length L 'in infinitum' and join up the dots (X,Y co-ordinates) you produce a curved line thereby demonstrating that a centripetal force is continually acting on the body in the direction of the force centre and the triangular areas will always be proportional to the time passed by the body traversing each triangle; **QED**

Newton's constant of motion 'h' is not to be confused with the perpendicular distance 'h' shown in the above diagram; they are neither the same nor in any way connected.

Newton's diagram is less easily seen from his words and his fairly simple diagram than if you actually complete it, and repeat it for ever smaller values

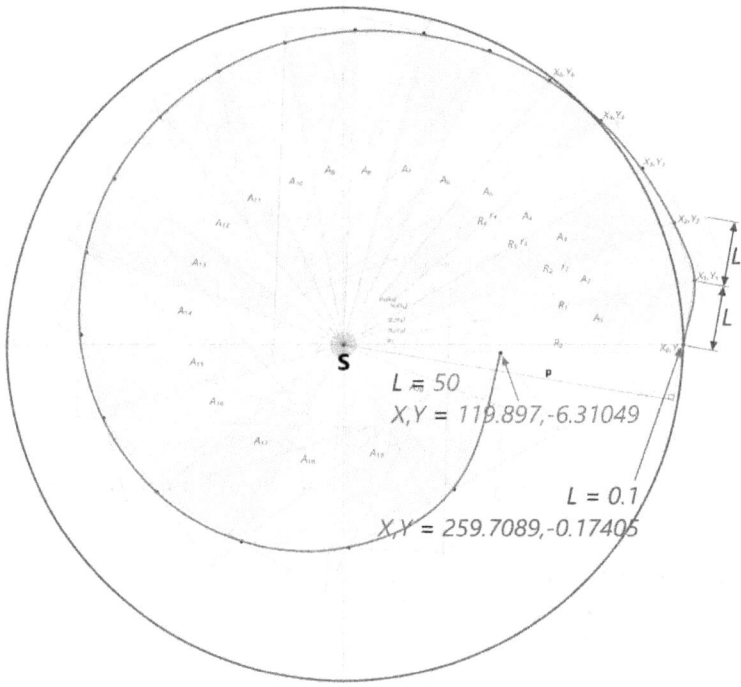

of L through to 360°:

A calculation was carried out using the following input data:

$X_o, Y_o = $ **260,0**

$L = 50$

$\theta = 100°$

The Mathematical Laws of Natural Science

As can be seen above, Newton's diagram does indeed produce a curve, exactly as he claimed, and following this through a sequence of diminishing values for L from 50 to 0.1, the following X,(Y) co-ordinates are achieved immediately prior to reaching 360° ...

L	X	(Y)
50	119.74535317	(-5.41947224)
25	179.77612136	(-11.51597352)
10	229.5157957	(-9.468518876)
1	257.067565265	(-0.349111926)
0.1	259.707825986	(-0.076479042)

Note: the 'Y' co-ordinate is in parenthesis because it is simply a resultant. The trend is demonstrated by the 'X' co-ordinate.

... from which it isn't difficult to anticipate where X (& Y) will end up if L is diminished in infinitum [i.e.: X(,Y) = **260(,0)**], making the final shape a circle and thereby proving that:

a) the path of the body is continuous (conservation of energy and momentum)

b) the orbital time passed by the body is proportional to the swept area (triangle)

c) Newton's calculus can be used to determine the properties of the path {'in infinitum'}

This result does not mean that the orbital path is circular, simply that it is continuous.

Corollary 1

Newton's first Corollary (to the above proof) states that the velocity of the body (v), represented by L, at positions A, B, C, D, E, F, etc. (Newton's diagram) is inversely proportional to the perpendicular distance of its tangent from the force centre (p)

Newton also stated that; v multiplied by p is a constant, i.e. his constant of motion (h), which is the angular momentum without the *mass* component.

The Mathematical Laws of Natural Science

Using the above 'in infinitum' argument it can be seen in the following table where these calculations have been carried out for successively reduced values of L between the start and end of the orbit (h_o @ 0°, h < 360°), 'h' does indeed become a constant:

L	H_o	h
50	47.46494162	106.803277
25	24.20484775	35.53390597
10	9.782449005	11.14661354
1	0.984150369	0.996040786
0.1	0.098474198	0.098591562

The Mathematical Laws of Natural Science

A-5.6.2 Centripetal Force

Centrifugal acceleration (according to Christiaan Huygens {1629 to 1695}):

$\alpha = R.\omega^2$

where $\omega = 2.\pi/t$

$a = \sqrt{[(R.\omega^2)^2 + (R.\alpha)^2]}$

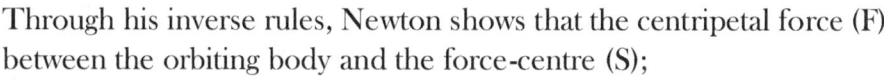

with constant angular momentum; $\alpha = 0$

$a = R.\omega^2 = R.(2.\pi/t)^2$

Centrifugal force:

$F = m.a = m.R.(2.\pi/t)^2 = 4.\pi^2.m\,(R/t^2)$

Through his inverse rules, Newton shows that the centripetal force (F) between the orbiting body and the force-centre (S);

$F = SP^2 . QT^2 / QR$

$PR = v^P . \delta t$

where; v^P is the velocity of the body at P and δt is the time taken for the body to travel from P to Q

$QR = (F^P / 2m).\delta t^2$

where; F^P is the centripetal force on the body at P

$F = QR . (2m / \delta t^2)$

where; F^P is the centripetal force on the body at P

$\delta t = PR/v^P = PR / (h/SY) = PR . SY / h$

where h is Newton's constant of motion (see Corollary 1 above)

Therefore, the centripetal force (F) can be calculated as follows:

$F = QR . (2.m / (PR . SY / h)^2)$

$\quad = QR . (2.m / (PR^2 . SY^2 / h^2))$

$\quad = QR . (2.m.h^2 / (PR^2 . SY^2))$

$\quad = QR.2.m.h^2 / PR^2.SY^2$

Newton preferred the calculation in geometric form by setting $2.m.h^2$ as a constant (k):

$F = k.QR / (PR.SY)^2$

The Mathematical Laws of Natural Science

A-5.6.3 Distance Between a Satellite & its Force-Centre (R)

The separation (distance) between an orbiting body and its force centre, can be found by using general elliptical equation:

$R = a.(1-e^2) / (1-e.\cos(\theta))$

Where; 'R' is the distance between the satellite and its force-centre at 'θ', 'θ' is the angle of the satellite's orbital progress from its apogee and 'e' is the orbital eccentricity

The force centre is not at the centre of an ellipse but at its focus (S)

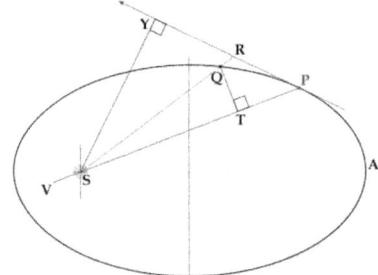

A-5.6.4 The Inverse Square Law

Proposition XI: "If a body revolves in an ellipse; it is required to find the law of the centripetal force tending to the focus of the ellipse"

Using similar geometric arguments as above (A 5.6.1 to 3) Newton worked out that the force between an orbiting body and its force centre is proportional to the inverse of their separation (the distance between them): $F \propto 1/R^2$ i.e. $F = K / R^2$

where: the constant of proportionality: $K = G.m_1.m_2$; i.e. $F = G.m_1.m_2 / R^2$

where G is a constant and m_1 and m_2 are the masses of the force centre and the orbiting body

This same relationship ($F \propto 1/R^2$) also applies to parabolas and hyperbolas as well as the ellipse

The above constant of proportionality (K) can also be written as; $K = m.h^2/p$

Where 'h' and 'p' are defined in Corollary 1 above and m is the mass of the orbiting body; i.e. $F = (m.h^2/p).(1/R^2)$

In the first formula, you can resolve the problem knowing the mass of the bodies

In the second formula, you can resolve it knowing the velocity and mass of the orbiting body and the parameter of its curve (p)

Both of the above F calculations produce the same result;

e.g. the following centripetal force occurs in the earth's orbit, 0.000175° from the major semi-axis:

G = 6.67359232004332E-11 (gravitational constant)
m_1 = 1.9885E+30 (sun mass)
m_2 = 5.964519768E+24 (earth mass)
R = 1.5209420E+11 (distance between mass centres)
$F = G.m_1.m_2 / R^2$ = **3.421649078E+22** (centripetal)
h = 4.454920463E+15 (constant of motion - see Corollary 1 above)
m = 5.964519768E+24 (earth mass)
p = 1.495528319E+11 (ellipse parameter)
R = 1.5209420E+11 (distance between mass centres)
$F = m.h^2 / p.R^2$ = **3.421649078E+22**

A-5.6.5 Orbital Period

Proposition XV: "The same things being supposed, I say, that the periodic times in ellipses are as the $^3/_2{}^{th}$ power of their greater axes"

This means that if the major semi-axis of an ellipse is 'a' (A 5.5) and the time taken for a body to orbit the elliptical path is 't' then the relationship between the two is:

$t \propto (2.a)^{1.5}$ or $t^2 \propto (2.a)^3$

Therefore; $t = K \cdot a^{1.5}$

Where K is the constant of proportionality, which is dependent on the properties of the force-centre.

This is actually Kepler's third law

A-5.6.6 Constant of Proportionality

To determine 'K' (the constant of proportionality for $t = K \cdot a^{1.5}$)

...

$K = t^2/a^3 \ \{s^2/m^3\}$

now we know ...

... that the earth travels around the sun in 31558149s

... the earth's semi-major orbital x-axis is 1.495945981E+11m

Therefore:

$t^2/a^3 =$ **2.974914364E-19** $\{s^2/m^3\}$

$G = 6.67359232\text{E-}11 \ \{N.m^2/kg^2 = kg.m.m^2 / s^2.kg^2 = m^3 / s^2.kg\}$

$m_1 = 1.9885\text{E}+30$ kg (the *mass* of our sun)

$1 / m_1.G = 7.535546116\text{E-}21 \ \{s^2.kg / m^3/kg = s^2/m^3\}$

2.975944645E-19 ÷ 7.538155846E-21 = 39.47841760436

$\sqrt{39.47841760436} = 6.2831853071796 = 2.\pi$

Therefore:

$K = (2\pi)^2 / G.m_1 = (2\pi)^2 \div 6.67359232\text{E-}11 \div 1.9885\text{E}+30$

   = **2.974914364E-19** s^2/m^3

i.e.;

$K = (2\pi)^2 / G.m^{fc}$

where m^{fc} is the *mass* of the force-centre

The above calculation, based upon NASA's data for the sun and the earth's orbit, gives an error margin of 0

The Mathematical Laws of Natural Science

A-5.6.7 Alternative Velocity Calculation

A much simpler orbital velocity calculation method is based upon Kepler's 'swept-area = time' rule

Using the earth's orbit as an example:

Earth's total orbital area (A) is 7.029445371E+22m² and it takes 31558149s (t) to complete

The swept area (A) is equal to ½.R x 1m

The velocity of the orbiting body at any given distance between the centres of *mass* (P & S) is calculated as follows:

$v = 2.A / t.R \ \{m^2 / s.m = m/s\}$

By way of verification:

The earth's maximum velocity occurs when R = 1.47095E+11 m (@ A)

$v = (2 \times 7.029445371E+22) \div (31558118.4 \times 1.47095E+11)$

   = **30286.008788376** m/s

30286.008788376 m/s (calculated using; h = v.R)

The earth's minimum velocity occurs when R = 1.520941962E+11 m (@ B)

$v = (2 \times 7.02944537126484E+22) \div (31558149 \times 1.520941962E+11)$

   = **29290.5355716777** m/s

29290.5355716777 m/s (calculated using; h=v.R)

The above confirms Kepler's 'swept-area = time' rule and shows that

$v \propto 1/R$

or $v = k/R$

where $k = 2.A / t$

A-5.6.8 Centrifugal force in an orbiting body

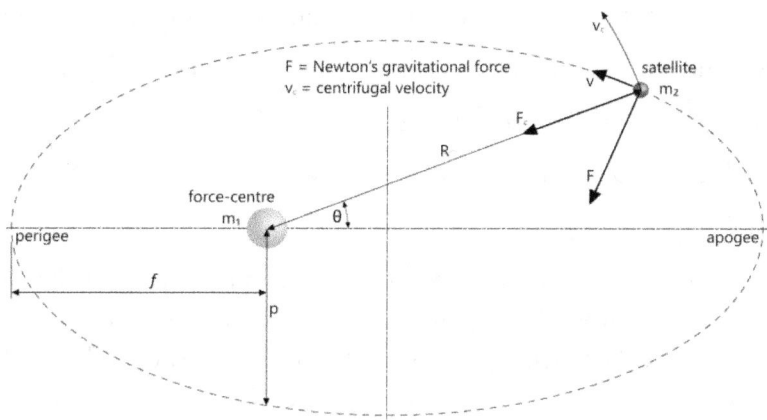

In any orbiting system, the centripetal force, i.e. Newton's *gravitational* force, must be equal to the orbiting body's centrifugal force, which may be calculated thus:

$F = m_1.v_2^2 / R$

$Fc = m_1.v_2^2 / R$

where

$v_2 = \sqrt{[G.m_1 / R]}$

@ the perihelion (perigee) of an ellipse; $Fc = F . f/p = F/(1+e)$

@ the aphelion (apogee) of an ellipse; $Fc = F . p/f = F.(1+e)$

Orbital velocity anywhere in an orbit may be calculated thus:

$v_2 = 2\pi.a.b / R.t$

where: t is the satellite's orbital period

Centrifugal velocity anywhere in an orbit may be calculated thus:

$\alpha = \sqrt{[4/3.\pi]}$

$\zeta = \sqrt{[(f.\text{Sin}(\theta/2)^\alpha + p.\text{Cos}(\theta/2)^\alpha) / (f.\cos(\theta/2)^\alpha + p.\text{Sin}(\theta/2)^\alpha)]}$

$v_c = \zeta.v_2$

The Mathematical Laws of Natural Science

A-5.6.9 Fundamental Laws of Orbital Motion

1) Every orbital system must have a force-centre and at least one satellite

2) A force-centre's *mass* defines its satellite's orbital shapes and periods

3) Satellite orbits define a force-centre's spin

4) Sub-satellite orbits and force-centre spin define a satellite's spin

5) Sub-satellites have no effect on the force-centre

6) Satellites may be swapped between orbits without altering orbital shapes and periods; e.g. Jupiter may replace Earth and Jupiter would follow the same orbital path that Earth previously followed and would orbit in 365¼ days

The Mathematical Laws of Natural Science

A-6 Newton Atom vs Planck Atom

Planck's particles were used to establish 'Σ' which was used to establish Newton's gravitational constant; 'G'.

Newton's atom is the real one; the one we see, hear and feel all around us. Planck's atom is theoretical, it does not exist. It is based upon his three constants (Table A 6.1A below); time (t^p), length (λ^p) and mass (m^p).

It should be noted here that Planck's length is actually an orbital radius and not a wavelength (λ). The reason being; force (F) is Energy divided by distance, and a distance in such a formula can only be an orbital radius.

The atomic particles in Planck's atom are identical in size; his electron is exactly the same, in all respects, as his proton. The only consistency between the two atoms (Newton's and Planck's) is that the product of the particle volumes in each atom is identical; $V_e.V_p = V^p.V^p$

It is the comparison between these two atoms that has given us the ability to solve Newton's gravitational constant

Tables A6.1A to C provide comparative results for three atomic variations in which the particle properties are defined thus:

Newton atom; comprising Newtonian particles (Table A 6.1A)
$t = a_o/c$; $\lambda = a_o$; $m = \rho_u/1m^3$; $E = \rho_u.c^2$; $F = E/a_o$

Planck^N atom; comprising Newtonian particles (Table A 6.1B)
t^p; λ^p; m^p; E^p; F^p (Table A 6.1B)

Planck^P atom; comprising Planck particles (Table A 6.1C), and calculated using a revised version of his own constant; h: h' = ½.$R_n.m_e.c^2$ J.m
Note: 'c', 'ε₀', 'G', 'φ' and 'k' are equal in both Newton and Planck atoms

The solution:

$F^N = G.m_e.m_p/a_o^2$ $F^P = G.c^4$

$\Sigma = V_e.V_p = V^{P2} = 3E-91$ (m^6) $\Sigma = F^N/F^P = 3E-91$ (no units)

$G = \sqrt{[\Sigma.a_o^2.c^4 / m_p.m_e]} = 6.67359232004332E-11$ m³/kg/s²

$G = a_o.c^2/m_u = 6.67359232004332E-11$ m³/kg/s²
where; m_u = one cubic metre of ultimate density matter (one Newton mass)

The Mathematical Laws of Natural Science

Newton atom compared with a Planck^N atom:

	Newton Atom (A)	Planck^N Atom (B)	A/B
t	1.76514516887831E-19	5.39096122598359E-44	3.27426797353056E+24
λ	5.2917721067E-11	1.61616952231128E-35	3.27426797353056E+24
m	7.1266079635045E+16	2.1765500017459E-08	3.27426797353056E+24
E	6.40507585675678E+33	1.95618559889902E+09	3.27426797353056E+24
F	1.21038391820525E+44	1.21038391820525E+44	1

Table A 6.1A

Newton atom compared with a Planck^P atom:

	Newton Atom (A)	Planck^P Atom (B)	A/B
t	1.76514516887831E-19	1.48432887846076E-34	1.18918737922067E+15
λ	5.2917721067E-11	4.44990604438464E-26	1.18918737922067E+15
m	7.1266079635045E+16	59.9283854507006	1.18918737922067E+15
E	6.40507585675678E+33	5.38609471364748E+18	1.18918737922067E+15
F	1.21038391820525E+44	1.21038391820525E+44	1

Table A 6.1B

Planck^N atom compared with a Planck^P atom:

	Planck^N Atom (A)	Planck^P Atom (B)	A/B
t	5.39096122598359E-44	1.48432887846076E-34	2.75336589568949E+09
λ	1.61616952231128E-35	4.44990604438464E-26	2.75336589568949E+09
m	2.1765500017459E-08	59.9283854507006	2.75336589568949E+09
E	1.95618559889902E+09	5.38609471364748E+18	2.75336589568949E+09
F	1.21038391820525E+44	1.21038391820525E+44	1

Table A 6.1C

Note: 3.27426797353056E+24 ÷ 1.18918737922067E+15 = 2.75336589568949E+09

The Mathematical Laws of Natural Science

The following Table provides the properties of particles in the atom that exists in nature; Newtonian particles.

Symbol	Formula	Value	Units
m_e		9.1093897E-31	kg
The *mass* of an electron			
m_p	$m_e \cdot \xi_m$	1.67262163783000E-27	kg
The *mass* of a proton			
m_n	$m_e + m_p$	1.6735325768E-27	kg
The *mass* of a neutron [4]			
V_e	m_e / ρ_u	1.27822236702922E-47	m^3
The volume of an electron			
V_p	m_p / ρ_u	2.34700946985653E-44	m^3
The volume of a proton			
V_n	m_n / ρ_u	2.34828769222356E-44	m^3
The volume of a neutron			
r_e	$\sqrt[3]{3 V_e / 4\pi}$	1.45046059426276E-16	m
The radius of an electron			
r_p	$\sqrt[3]{3 V_p / 4\pi}$	1.77613270336827E-15	m
The radius of a proton			
r_n	$\sqrt[3]{3 V_n / 4\pi}$	1.77645508248591E-15	m
The radius of a neutron			
t^N	a_o / c	1.765145168878310E-19	s
Newton time			
λ^N	a_o	5.291772106700000E-11	m
Newton length			
m_u	$a_o \cdot c^2 / G$	7.126607963504500E+16	kg
Newton *mass*			
E^P	$m_u \cdot c^2$	6.40507856756780E+33	J
Newton energy			
F^N	E^P / λ^N	3.63115175461574E-47	N
Newton force			
ρ_u	$\sqrt{[m_e \cdot m_p]} / \sqrt{\Sigma}$	7.1266079635045E+16	kg/m^3
The ultimate density of Newton's atomic particles			

Table A 6.2

The Mathematical Laws of Natural Science

The following Table provides the properties of particles in a fictitious atom that has been constructed using Planck's values; t^P, λ^P, m^P

Symbol	Formula	Value	Units
m_e^P	m^P	2.1765500017459E-08	kg
Planck's electron *mass*, which is equal to Planck's mass			
m_p^P	m^P	2.1765500017459E-08	kg
Planck's proton *mass*, which is equal to Planck's *mass* (and Planck's electron)			
m_n^P	N/A	N/A	N/A
Planck's neutron is unnecessary as his atom is theoretical only			
V_e^P	$\sqrt{\Sigma}$	5.477225575051 66E-46	m^3
Planck's electron volume			
V_p^P	$\sqrt{\Sigma}$	5.477225575051 66E-46	m^3
Planck's proton volume			
V_n^P	N/A	N/A	N/A
Planck's neutron is unnecessary as his atom is theoretical only			
r_e^P	$\sqrt{[\,3.V_e^P / 4\pi\,]}$	5.07563837996471E-16	m
Planck's electron radius			
r_p^P	$\sqrt{[\,3.V_p^P / 4\pi\,]}$	5.07563837996471E-16	m
Planck's proton radius			
r_n^P	N/A	N/A	N/A
Planck's neutron is unnecessary as his atom is theoretical only			
t^P	$\sqrt{[\,\hbar.G / c^5\,]}$	5.39096122598359E-44	s
Planck time			
λ^P	$\sqrt{[\,\hbar.G / c^3\,]}$	1.61616952231128E-35	m
Planck length			
m^P	$\sqrt{[\,\hbar.c / G\,]}$	2.1765500017459E-08	kg
Planck mass			
E^P	$\sqrt{[\hbar.c^5 / G]} = m^P.c^2$	1.95618559889902E+09	J
Planck energy			
F^P	c^4 / G	1.21038391820525E+44	N
Planck force			
ρ_u^P	m^P / V_e^P	3.97381844498046E+37	kg/m^3
The ultimate density of Planck's atomic particles			

Table A 6.3

A-7 What Went Wrong?

The 17th century saw the start of the industrial revolution, which has since become a technological revolution; but it's no different. Its growth remains today dependent on guesswork (experimentation).

The beginning of the 20th century should have seen the start of a scientific revolution, enabling all technological development to be achieved mathematically.

It never happened. Why?

The scientific community were successfully intimidated by a couple of individuals in the first quarter of the twentieth century whose theories were plainly wrong.

It subsequently became acceptable practice to invent bizarre irreconcilable theories claiming that "*the known laws of physics do not apply*" is acceptable justification. In fact, all of today's theories not only disobey the known laws of physics, they rarely relate with each other.

The entire scientific community has become so entrenched with dogma associated with these theories, that it is now impossible to admit that it may be wrong without losing face, and just as with the religious community it displaced, bigotry and arrogance rules everything within.

Over a hundred years later, technological development continues to be suppressed because scientific stagnation cannot free it from guesswork.

All of today's branches and sub-branches of scientific study have developed to accommodate these problems, along with the inevitable search for numerous unification theories.

Science today is in a mess.

So, what went wrong?

Einstein and Bohr ...

A-7.1 The Error

The problem came about because of a 20th century misinterpretation of a 19th century experimental discovery.

It is currently believed that electrons emit light (*photons*); **they don't.**

We have been taught this for a hundred years, forcing us to create fanciful theories to explain how *mass* moves in waves; **it doesn't.**

The *photon* exists in our minds because of a very simple mistake made a long time ago relating to Crooke's tube.

Crooke, and everyone since, believed that he had created a perfect vacuum by pumping out all the air from his tube; **he hadn't!** His tube contained a *measured* vacuum, it was not a *true* vacuum; there were billions of protons inside it. It is impossible to create a perfect vacuum on planet Earth, or anywhere else on, or in, a celestial body where all matter resides.

When Crooke fired electrons from one end of his tube to the other, they *appeared* to emit light. So, he and everyone since believed that electrons must emit light. But the light you are seeing is not emitted by electrons, it is the electro-magnetic radiation emitted due to their interaction with protons in the tube. When a bar magnet is placed beside the tube, the light path deflects. What you see is not the bending of light (although magnetism does [slightly] bend light; the dramatic deflection you see is that of the *path* of the electrons; the light is emitted by interactions along this deflected path.

During his 'light-on-a-metal-plate' experiment Max Planck detected a feint but perceptible electric charge that '*pulsated*, confirming that it was induced by electro-magnetic energy. If it had been induced by a stream of electrons (*photons*) it would have been continuous (DC) as in a battery, confirming that light is electro-magnetic energy (AC) not electrons. The light emitted by stars and everything else in the universe is not brought to us by electrons: it is radiated electro-magnetic energy. Michael Faraday understood this and James Clerk Maxwell described it mathematically.

These are the main reasons why Relativity and Quantum Theory remain unproven after 100 years; they both rely on photons.

The Mathematical Laws of Natural Science

A-7.1.1 Measured Vacuum

A *measured* vacuum defines the number of protons inside Crooke's Tube:

Crooke would have had to remove every proton from inside his tube to claim that electrons fired within it were emitting light.

His tube was originally filled with air: 78% nitrogen, 21% oxygen & 1% argon.

This plot shows the number of protons inside his tube at pressures between 7.5E-18 bar and 1E-05 bar.

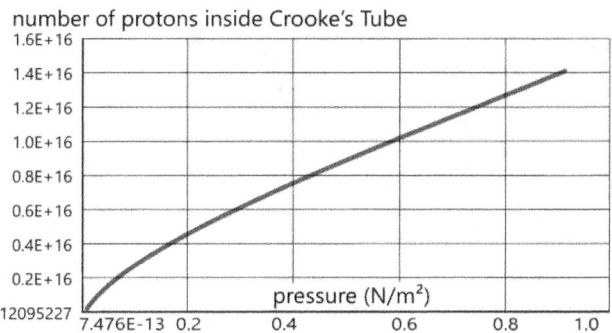

Gas pressure, including that in air, originates from the force generated between adjacent atoms by their repulsive electrical-charge.

This repulsion (force) drops off very quickly with the [square of the] distance between the atoms (exponential curve), and as can be seen from the above graph, even at 7.5E-13 N/m² (7.5E-018 bar), there would still have been about 12 million protons inside his tube.

Crooke used a mechanical pump to remove, what he thought was, all the air from inside it.

However, today's best laboratory mechanical pumps could achieve little better than 1E-10 bar (1E-05 N/mm²), which relates to about 3E+12 protons left inside his tube. It is expected that Crooke actually achieved about 1E-05 bar, which meant there were about 1.4E+16 protons left inside his tube.

A-7.2 The Problems with Relativity

For reasons of brevity, I shall refer to the theories of general and special relativity as 'Relativity' and the author of these theories as the 'Author'.

This chapter includes numerous reasons why Relativity can no longer be considered appropriate for orbital motion, the most significant of which are:

1) *All* orbits work perfectly well without it (i.e. it is unnecessary).

2) It causes many aspects of Newton's laws of orbital motion to fail.

3) It was developed simply to support item 4) below.

4) The Author misunderstood the meaning of both $E=mc^2$ & light.

Just one, or even two, of the following problems (A10.2.1 to .8) could be chalked up to coincidence, but all eight!

A major concern regarding Relativity is the lack of attention paid to matching units in its formulae. For example, it appears to include the formula: '$R_{ab} - \frac{1}{2}R.g_{ab} = T_{ab}$'; in which length is added to velocity squared which results in time. Even based upon Reimann mathematics, this doesn't make much sense.

It is important to remember:

Whilst it is possible to create a sub-theory to explain any distortion of reality you wish, why would you if there is no need?

When everything in the universe can be explained without the sub-theory, the sub-theory becomes redundant.

Relativity was driven by a desire to explain events that were either unknown or misunderstood. Now that we fully understand the theory behind *all* orbital systems and *light*, Relativity has become redundant, especially as it actually invalidates Newton's laws of orbital motion, that otherwise work perfectly well, *in every respect*, irrespective of energy, speed and *mass*.

It seems clear to me, that Relativity must be declared '*dead in the water*'

The Mathematical Laws of Natural Science

A-7.2.1 The Speed of Light

All physicists currently claim that the light we see is emitted by *photons*.

They also claim that electrons are 'weird beasts' that possess *mass* and travel in waves, which is the reason we cannot pin them down (uncertainty principle).

This is difficult to understand given that; if the entire electro-magnetic spectrum ranges between <2E-14m and >7m, how can all *photons* travel at the same velocity. Surely, they must travel at all speeds between >0 to 'c' in order to represent the full electro-magnetic spectrum.

For example; an electron travelling at 1E+06m/s will possess a different energy to one travelling at 1E+08m/s, and if electrons emitting different colours, for example, travel at different velocities they can't all be travelling at 'c', so they cannot be *photons*!

Moreover, according to Newton's laws of *gravitational* attraction, electrons travelling at different velocities must be deflected at different angles (refer to Appendix A-7.2.2: α), which would result in light passing our sun being separated into the various spectral colours, which doesn't happen.

Both theories of Relativity are based upon the fundamental principle that: "light possesses *mass*" and all wavelengths of light passing the same celestial body at the same radial distance is deflected at the same angle (α), both of which are contradictory.

Therefore, the light we see cannot be *photons*, it must be electro-magnetic energy, which possesses no *mass* and is deflected by magnetism at the same angle (α).

Not only is it unnecessary to deform space-time around celestial bodies in order to explain light deflection, it is mathematically incorrect to do so.

Special relativity was devised because of the inability to correlate the additive nature of *mass*-velocity with the non-additive nature of light. This is only a dilemma if light possesses *mass*, which it doesn't. The *photon* is therefore, also based upon a misunderstanding of the nature of light.

There is no such thing as a *photon*.

The Mathematical Laws of Natural Science

A-7.2.2 Light Deflection

Light is apparently observed to deflect by an angle of 1.75 arc-seconds when passing at or close to the surface of our sun (chapter 2.5.10; EME Deflection).

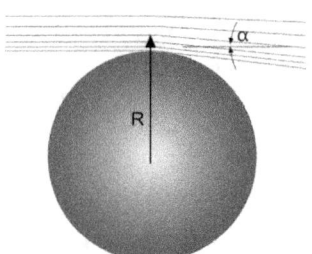

According to the Author, however, Isaac Newton's laws of *gravity* predict a deflection angle of half this value (0.875)

Relativity is a mathematical description of space-time/*gravity* distortion around celestial bodies that was developed to support the Author's formula for *observed* light deflection angles, which was based upon '*light-emitting photons*' and their susceptibility to *gravity*.

The problem with the Author's approach is that the light we see all around us is electro-magnetic energy, and therefore possesses no *mass*, so *gravitational* laws don't apply (refer to chapter 4.3).

On the other hand, Isaac Newton's *gravitational* constant (G), which is based upon the properties of Quanta, may be used to define this deflection angle (α) as follows:

$\alpha = \text{Atan}(4 \cdot m_s/m_u \cdot a_o/R)$ {~~kg.m~~ / ~~kg.m~~}[#]

Where:

$G = a_o \cdot c^2/m_u$ {m³ / kg.s²}

m_s = the *mass* of our sun {kg}

m_u = unit *mass* of ultimate density {kg/m³}

a_o = Rydberg's radius {m}

c = the speed of light in a vacuum {m/s}

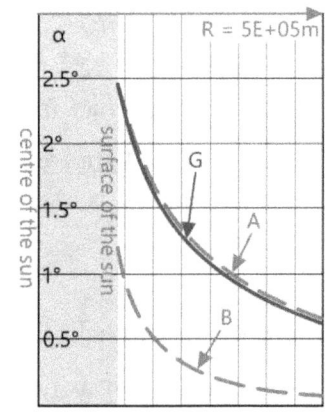

When calculating light deflection angles, the Author reduced the size of our sun by a factor of 5000, retaining its correct *mass* and increasing its density accordingly. Presumably, this was done to raise the calculated angles to a practical value.

Curve **A** shows the variation in 'α' according to **Relativity**, from the surface of the above modified sun to a distance 5E+05m from its centre, which complies with the *observed* deflection value.

Curve **B** shows the equivalent variation in 'α', according to the Author, when using **Newton's laws of *gravity*** on a '*photon*'.

Curve **G**; we can also reproduce the *observed* light deflection angles if we plot the light deflection angles using formula # (above), which is based upon **Newton's *gravitational* constant**. The difference being that this formula, which is not based upon the potential susceptibility of '*photons*', does not need a sub-theory to justify it.

It is important to understand that it isn't necessary to prove the validity of an alternative theory in order to discredit the original. You only need to demonstrate that the same result may be achieved by applying valid input data to an identical model but without the need for a sub-theory for justification. The above formula does exactly that.

That light travels in waves and not as particles is not new; Christiaan Huygens declared this to be the case in the late 17th century and it was later restated by Faraday, Maxwell and Pauling.

Newton on the other hand declared light to be particles. The Author automatically declared Newton to be correct because he quite rightly held Newton in such high esteem, and therefore used Newton's laws of *gravity* to deflect [particle] light as it passed celestial bodies. The problem was that this produced incorrect values (Curve **B**), but the Author needed to create a sub-theory (Relativity) to justify his approach.

General relativity, the deformation of space-time, is based upon the inability to use Newton's laws of *gravity* to predict the deflection of light, which is only a problem if light possesses *mass*, which it doesn't. This theory is therefore based upon a misunderstanding of the nature of light.

General relativity was also devised because of the Author's disbelief in force-fields, which he called the 'ether'. But whilst there is no such thing as *the ether* (as he understood it) we do know that force fields exist, as anybody holding two magnets close together, but not touching, will know.

A-7.2.3 Neutronic Radius (R_n)

The neutronic radius (refer to chapter 15.1), which is achieved by an orbiting electron when travelling at 'c', can *only* be explained using Newton's laws of orbital motion and Coulomb's law of electrical force. It occurs in far too

The Mathematical Laws of Natural Science

many constants (magnetic, permittivity, Rydberg's, Planck's, Coulomb's, Henry's, etc.) to be rejected as a *fundamental physical constant*.

The neutronic radius is also the basis of $E=mc^2$ (refer to chapter 2.5.12)

It cannot be a coincidence that Newton's force formula predicted it 300 years ago!

$$R_n = G.m_p / \varphi.c^2$$

The conversion of *mass* to energy with velocity together with the space-time/potential distortion around force-centres as defined in Relativity, would render such an orbital radius impossible. I.e. the electron would be orbiting inside the proton at 'c' and R_n would be incorrect, making magnetic constant, permittivity, Rydberg, Planck, Coulomb, Henry, etc. incorrect, which we know is not the case.

A-7.2.4 Station-Keeping

Relativity is based upon the predication that light possesses *mass* and that *gravity* is responsible for its deflection, and because the Author claimed that Newton's laws of potential attraction cannot apply to light, it was necessary to deform space-time around celestial bodies and artificially modify satellite velocity to account for this problem. But if we apply the relativistic velocity modification to a satellite, such as the earth;
$v = v / \sqrt{[1+(v/c)^2]}$ we find that; whilst Newton's laws <u>always</u> work (centrifugal and potential acceleration always balance) irrespective of satellite velocity, Relativity begins to fail well below 1% of *light-speed*. In fact, according to Relativity an electron can never actually achieve *light-speed*: $c \neq c/\sqrt{2}$

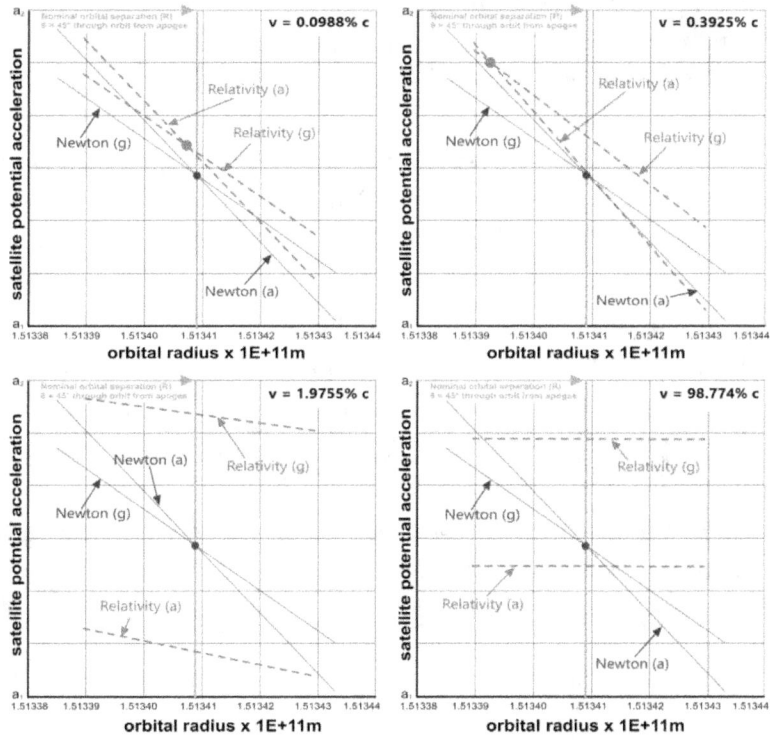

The above plots show the same calculation performed for an electron passing the sun at the *speed of light* at an orbital radius of 765061000min which Newton's laws of orbital motion function correctly with no modification to elliptical orbits or satellite velocity. Relativism, however, shows potential acceleration at *light-speed* is <u>always</u> greater (more than twice) centrifugal acceleration, meaning that a *photon* cannot pass the sun without

The Mathematical Laws of Natural Science

being absorbed by it if light possesses *mass* and *gravity* is responsible for its deflection (as the Author claimed).

In fact, simply because the Author assumed that light possesses *mass*, he found it necessary not only to modify space and time but also artificially modify velocity, and even then, his theory fails in that it predicts an electron cannot achieve *light-speed*, destroying his concept of a *photon*.

Elliptical orbits are an indisputable fact of nature. This has been repeatedly demonstrated since Kepler's discovery in the 17th century. Its mathematical laws show that an exact ellipse is *fundamental* to the constants - such as K, h, E, etc. - of [orbital] motion and thereby essential to maintain satellite paths in non-circular orbits. Relativity requires a distortion of this ellipse, rendering the orbital laws unworkable; yet we know that Newton's universal orbits work perfectly irrespective of size, shape and speed.

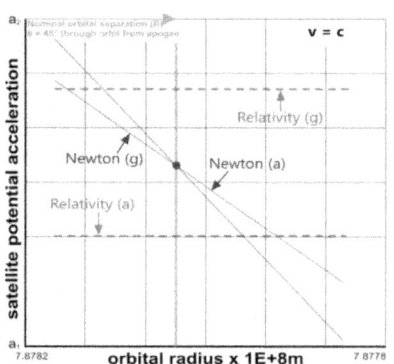

Moreover, if we alter time according to Relativity ($T_{AB} = R_{AB} - \frac{1}{2}.g_{AB}.R^{\#}$) the balance will shift ensuring that external interaction (from other bodies) cannot be rectified naturally; yet they do, as demonstrated in chapter 9.6.

\# *The units within this formula, which do not match, cannot be reconciled without a sub-theory. Such an anomaly does not occur in any of Isaac Newton's theories.*

It must therefore be concluded that Relativity is not only unnecessary; this aspect of the theory is also incorrect.

A-7.2.5 E=mc²

$E=mc^2$ applies to potential energy, not kinetic energy (refer to chapter 2.5.12).

The author therefore inappropriately applied Henri Poincaré's formula to "*photons*", which do not and cannot possibly exist.

The author's entire philosophy (Relativity) is based upon his belief that light possessed *mass* and that Henri Poincaré's formula ($E=mc^2$) refers to kinetic energy and implies that *mass* and energy transform with velocity.

He needed to develop numerous sub-theories such as the deformation of space and time and also claim that $c = c/\sqrt{2}$ to justify his philosophy, none of which can be verified using the known laws of physics and all of which remain unproven today.

A-7.2.6 Hades

Hades is a cold (dark) body just as is the case for all galactic force-centres.

At the time Relativity was theorised, neither its Author or anyone else was aware of the exigency of force-centres in *every* orbital system or of [planetary] spin theory (refer to chapter 10). The Author therefore misunderstood the effect of galactic population on orbital shapes (refer to chapter 9); hence the misguided inventions of black-holes and dark matter (refer to chapter 9.3.2).

Moreover, if the Author had known of Hades and the laws of station-keeping (refer to chapter 9.4), he would have realised that the deformation of space-time could not work.

It has been believed for some time now that black holes exist; **they don't**. They can't possibly exist because they do not obey any of the laws of; physics, energy or thermodynamics. Black-holes are simply large, cold, dark bodies most of which are galactic force-centres. They don't need to be large enough to trap *photons* because *photons* don't exist, they can be any size.

Because nobody understood what they were, weird and wonderful theories have abounded around them; one of which is the event-horizon ....

A-7.2.7 Micro-Lensing (Event Horizon)

Another discovery attributed to the Author is the Event-Horizon, which was used to explain halos that appear around some stars, but which:

1) was not discovered by the Author, it was proposed by Orest Khvolson and Frantisek Link; and,

2) is based upon a fallacy; Black-Holes; and,

3) can be explained simply by using Newton's *gravitational* constant and electro-magnetic energy to describe micro-lensing.

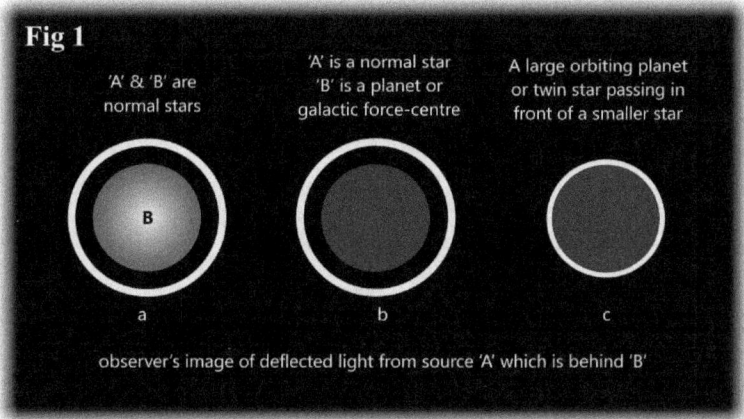

Fig 1 shows a number of lensing images as they might appear to an observer:

a): lensing that would be expected when the two stars concerned (a source star ('A') behind an observed star ('B')) are widely separated;

b): the same lensing condition as described for a) but the observed star is dark; either a galactic force-centre or a star with no planets.

c): the same lensing condition as described for a) but the bodies are much closer together, one of which is in orbit around the other - one will be dark and the other bright. The view shown here is when the dark body is directly between the bright body and the observer.

To explain:

The Mathematical Laws of Natural Science

We shall refer to your eye as the observer, the source as star 'A' and the intermediate (or blocking) body as star 'B' (Fig 2).

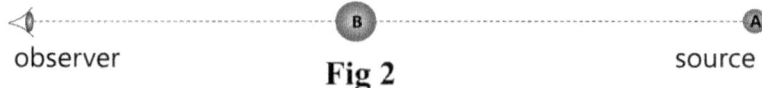

Fig 2

When passing a large (celestial) body (B), electro-magnetic energy from a source (A) is deflected by the magnetic charge in 'B'. The angle at which the light is deflected (α) may be calculated using Isaac Newton's gravitational constant.

Light from 'A' is radiated in all directions. If it wasn't blocked by 'B', it would appear to the observer as a normal star because all of the light that reaches the observer will travel in a straight-line, just as occurs with the light from 'B'. However, you would not expect to see 'A' because it is hidden behind 'B',

Fig 3

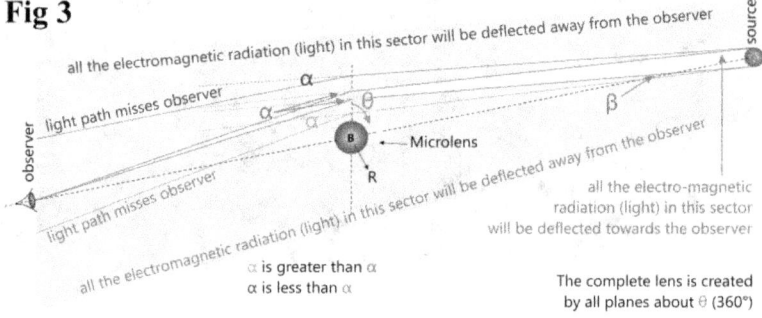

which you will see exactly as it is; a circular body.

If the angle of the electro-magnetic energy radiated from 'A' is just right (Fig 3; β), the deflected angle (α) will cause the electro-magnetic radiation to be directed at the observer. All the electro-magnetic radiation that is not emitted by 'A' at angle 'β' will not reach the observer. It will be deflected away from (below or above in the plane shown) the observer's location. Therefore, the region inside and outside the ring will appear black (Fig 1; a). The reason a micro-lens anulus has a thickness is due to the diameter of 'A', as can be seen in Fig 2.

If 'B' is a dead star or galactic force-centre, the inside of the micro-lens will appear all black (Fig 1; b).

The Mathematical Laws of Natural Science

Another manifestation of this phenomenon is for a smaller, hidden star to appear larger than the star in front of it. Exactly the same light deflection is taking place here but the bodies concerned are much closer together; e.g.

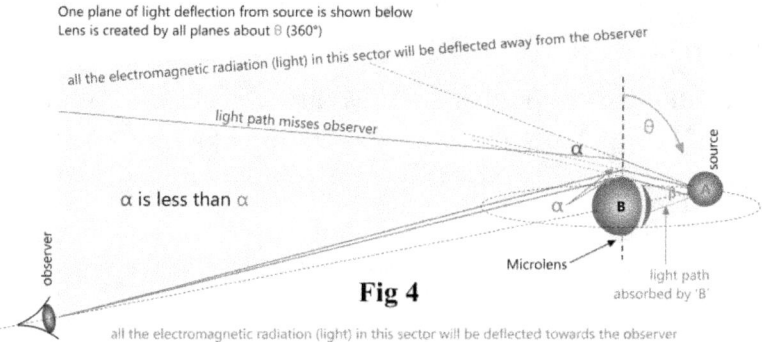

Fig 4

one of them is in orbit around the other (Fig 4).

Lensing created by a small, bright body passing behind a large star making a star appear larger than it really is to an observer (Fig 1; c).

In this case, 'R' will be significantly less than above (Fig 3) and the deflection much greater. And whilst the same lens will appear around 'B', it will appear at its surface, giving the appearance of a larger star behind it. In this orbital situation, one of the stars concerned is invariably dark, which is due to one of the bodies being a large planet, a dark star or a galactic force-centre.

Conclusion: There is no such thing as an event horizon because it is based upon a fallacy and it is unnecessary.

A-7.2.8 PVRT

The absolute nail in the coffin for Relativity is that the universally known and accepted formula for the pressure of gases; $P.V = n.R_i.T$ (ideal gas law) can be replaced with an alternative formula using the potential energy of a proton-electron pair (refer to chapter 8.4.2).

Relativistic velocities cannot predict gas pressures, the electron would be orbiting inside its proton partner as the temperature of a proton-electron pair approaches neutronic.

A-7.3 Quantum Theory

For reasons of brevity, I shall refer to Quantum Theory as 'QT' and its author as the 'Author'.

There are numerous reasons why QT can no longer be considered appropriate for the description of atoms, the most significant of which are listed below.

1) Whilst QT cannot explain or describe the behaviour of atoms in terms of what we see, feel and hear in the universe, an atom according to Isaac Newton and Coulomb can do this.

2) QT requires a still-undiscovered unification theory to ensure compliance with Newton's laws of motion.

3) The non-orbital nature of QT electrons means the QT atom cannot generate or emit electro-magnetic energy.

4) QT required the invention of 'string-theory' along with numerous sub-atomic particles (e.g. quarks, fermions, bosons, gluons, etc.) in order to make the atom work, whereas Newton and Coulomb can make the entire universe work with just two particles; the electron and the proton.

5) QT relies on statistics for justification; statistics apply only to the consequences of the laws of nature, never the laws themselves.

6) As is demonstrated in chapters 6, 7 & 8, Newton's and Coulomb's theories can be applied to the atom and the states of matter and therewith mathematically predict its properties.

7) QT's Author needed intimidation to force acceptance of his theories by the scientific community; "*if you aren't profoundly shocked by quantum physics, then you haven't understood it*".

8) QT remains unproven after 100 years.

9) It was necessary to invent sub-theories (including the uncertainty principle) to explain why electron location cannot be predicted in QT. This approach is similar to that devised by religious communities for their gods; "*I refuse to prove I exist says God, for proof denies faith and without faith I am nothing*". It is an untenable position.

10) That PVRT can be replaced with PE proves that electron's orbit protons, disqualifying QT, which is based upon non-orbiting electrons.

The Mathematical Laws of Natural Science

The single biggest problem with **QT** (item 3 above) is that its atom can only absorb energy; it has no way to emit it. However, items 6 and 10 above also proves that Newton's is the correct atomic model.

QT was driven by a desire to explain events that were either unknown or misunderstood. Now that we fully understand the theory behind all orbital systems and that light does not possess *mass*, **QT** has become redundant, especially as it does not obey Newton's laws of orbital motion.

Whilst it is possible to create a sub-theory to explain any distortion of reality you wish, why would you if there is no need?

When everything in the universe can be explained without a sub-theory, the sub-theory becomes redundant.

It seems clear that Quantum Theory must be declared '*dead in the water*', given that it fails to address the atom's single most important issue; the emission of electro-magnetic energy, whilst Newton's and Coulomb's laws together can explain all aspects of atomic structure and performance

A-8   The Molecule

I am not providing a solution for the molecule (see Epilogue below), this appendix is just a hint.

Electron clouding is responsible for holding atoms together in a viscous state. If electron clouding is high, atomic bonding (viscosity) is strong. As temperatures rise, shell radii decrease, reducing electron clouding and weakening inter-atomic bonding. However, molecular bonding persists even in gaseous state

All molecular bonding is ionic, as a result of ionic bonding. There is no such thing as covalent bonding. This mechanism is also responsible for *chemical* bonding between atoms.

Ionic bonding is the coupling of atoms due to ionic disparity (opposite electrical polarity in their nuclei). One atom will have a spare [lone-]proton within its nucleus and the other will have excess electrons.

In theory, any two atoms can form an ionic bond, but stable atomic bonds tend to be restricted to small atoms with low values (Γ):

Hydrogen (0.07146)	Magnesium (0.22875)
Helium (0.011709)	Aluminium (0.6795263)
Carbon (0.01605)	Silicon (0.0549643)
Nitrogen (0.0086143)	Phosphorus (0.5842566)
Oxygen (-0.0001125)	Sulphur (0.0365625)
Fluorine (0.998403)	Chlorine (0.7692353)
Neon (0.16173)	Potassium (0.5202474)
Sodium (0.8098118)	Calcium (0.0351)

The smaller the atom, the lower its 'Γ' value and the more stable its bond. Therefore, most molecules tend to include those above highlighted in **bold type**. Whilst sulphur, calcium & silicon all have low 'Γ' values, their size inhibits stability.

A low 'Γ' value signifies a low neutronic ratio (ψ). Because of the instability of molecules with high 'Γ' values it must be concluded that neutrons inhibit atomic bonding, which supports the premise for inherent stress in neutrons (refer to Chapter 5.3.5).

Helium is the odd-man-out here. Despite being small and having a low 'T' value, helium cannot form part of a natural molecule because it is extremely electrically stable; it is in fact, the most stable of all the atomic elements. This is because it has just two proton-electron pairs both of which orbit in shell-1, giving it the strongest possible electrical structure; it is perfectly balanced electrically, making it very difficult to share its electrical charges with other atoms.

The following natural bonds form between same-element atoms (diatomic) in gaseous form: Nitrogen (N_2), Oxygen (O_2), Fluorine (F_2), Chlorine (Cl_2), Bromine (Br_2), Iodine (I_2).

Note, whilst hydrogen is generally recognised as a diatomic atom; H_2 is an impossible molecule unless it in the form of deuterium or tritium.

Atomic size inhibits stable molecular bonding in Bromine & Iodine with anything but same-element atoms.

The lowest 'T' values can occasionally generate triple bonds ...

Carbon, Nitrogen, Oxygen (e.g. O_3)

... but they tend to be very unstable (e.g. ozone)

As long as you concentrate on Ionic strength, you will find the answer!

Epilogue

You will have noticed in this book, that the pre-twentieth century scientists had got it right, it simply needed somebody to put it all together.

The Newton-Coulomb approach to physics explains; every property of every particle, atom and matter, at any temperature and pressure, anywhere in the universe, using just one theory. It needs no fudging or surmise, and it is exact, repeatable and predictable.

You will also have noticed that I have not provided the solution to the molecule (appendix A-8), however, all the information to solve it is in this book, you just need to extract it. Whilst I now know how it works, I am leaving it up to you to complete because of the negativity I have experienced from all of our '*leaders*' along with their inability or unwillingness to disprove this work.

The academic community was sufficiently intimidated by Einstein and Bohr simply to accept their bizarre theories, neither of which are proven (or provable) after more than a hundred years of trying. This blind acceptance resulted in a proliferation of magical theories to back up their acceptance, none of which can be justified or realised. The scientific community is now so entrenched in magic that it has become impossible to put things right without losing face.

The discovery of a single mathematical theory that explains every aspect of nature, means that we can now (at last) replace today's pictorial approach to chemistry with mathematics, and eliminate our reliance on guesswork, unreliability, long-term testing and excessive costs associated with laboratory experimentation. It means we can *calculate* perfect materials, chemicals and medicines in seconds.

Whilst the entire content of this book may or may not be perfect, it is without doubt the most realistic, accurate *and verifiable* summary of our universe and the way it works published in any form to-date. It should not be difficult for the rest of humanity to finalise and correct where necessary.

This work, which took me just three years to complete, will enable the human race to obtain limitless clean free energy, virtually free transport and perfect materials, chemicals and medicines from a computer terminal in seconds

One day, somebody will have the will and the means to force our academic communities to lose face and accept the truth. I just hope it doesn't take another two-thousand years for this to happen.

THE MATHEMATICAL LAWS OF NATURAL SCIENCE

Cover illustration created by Eléonore Dixon-Roche